THE GHOSTS OF
IRAQ'S MARSHES

THE GHOSTS OF IRAQ'S MARSHES

A History of Conflict,
Tragedy, and Restoration

Steve
LONERGAN

Jassim
AL-ASADI

In collaboration with
Keith Holmes

The American University in Cairo Press

Cairo New York

First published in 2024 by
The American University in Cairo Press
113 Sharia Kasr el Aini, Cairo, Egypt
420 Lexington Avenue, Suite 1644, New York, NY 10170
www.aucpress.com

ISBN 978 1 649 03325 3

Library of Congress Cataloging-in-Publication Data
Names: Lonergan, Stephen C. (Stephen Colnon), 1950- author. | Al-Asadi, Jassim, 1957- author. | Holmes, Keith (Geospatial scientist), cartographer.
Title: The ghosts of Iraq's marshes : a history of conflict, tragedy, and restoration / Stephen Lonergan and Jassim al-Asadi ; in collaboration with Keith Holmes.
Identifiers: LCCN 2023003906 | ISBN 9781649033253 (hardback) | ISBN 9781649033260 (epub) | ISBN 9781649033277 (adobe pdf)
Subjects: LCSH: Al-Asadi, Jassim, 1957- | Marshes--Iraq--History. | Marsh conservation--Iraq--History. | Iraq--Environmental conditions.
Classification: LCC GB628.81.I72 L66 2023 | DDC 956.7/5 [B]--dc23/eng/20230928

1 2 3 4 5   28 27 26 25 24

Designed by Adam el-Sehemy

To the Marsh Arabs, heirs to the great Sumerian civilization.

إلى عرب الأهوار، ورثة الحضارة السومرية العظيمة.

TABLE OF CONTENTS

ACKNOWLEDGMENTS

I am grateful to many friends and colleagues for their valuable insights, suggestions, comments, and encouragement during the writing of this book. Some were familiar with the Marshes while others learned about the region for the first time. The Writer's Studio at Simon Fraser University was not only a welcome diversion during the months of COVID-19 isolation, but also provided a group of willing reviewers and editors for sections of the book. Brian Payton and JJ Lee, two writers associated with the studio, acted as mentors in the development of the manuscript. They encouraged me to use personal narratives to help readers better understand what happened in the Marshes, and their advice proved invaluable.

Hassan Partow and Ammar Al-Dujaili played an important role in reintroducing me to my co-author, Jassim Al-Asadi, and connecting me with other people and organizations working in the Marshes. Samira Abed Al-Shibeb and her staff at the Center for the Restoration of Iraqi Marshes and Wetlands provided useful insights into government policy and were open to speaking with me whenever I requested it. I am also grateful to Adel Al-Maajidy, his daughters, Asma and Assra, and Ihsan Kadhim Al-Asadi for speaking to me about their refugee experiences and being forced to leave the Marshes during the Shi'a Uprising in 1992. It was a pleasure meeting and talking to all of them.

I also benefited from the suggestions of numerous readers of the original draft of the manuscript. Doley Henderson, Erika Dickenson, Clive Blomer, Shelley Holmes, John Wheeler, John Talbot, and Gregg Sheehy, along with JJ Lee, reviewed, edited, corrected, and fact-checked the document, albeit sometimes questioning my sanity in undertaking the project. I

also appreciate the support of many friends who commented on individual chapters, despite their busy schedules. As always, my wife Arlene not only supported my writing efforts, but also read different versions of the manuscript and offered useful suggestions on the text and the photographs.
Steve Lonergan

FIGURES
(all figures are used with permission)

Figure 1. Location of the Mesopotamian Marshes.
Figure 2. Map of Iraq, with governorates (provinces).
Figure 3. The extent of the Marshes, 1973 and 2020.
Figure 4. West Hammar Marsh.
Figure 5. Traditional fishing in the Marshes.
Figure 6. Jassim, 17 years old.
Figure 7. Marsh village, circa 1975.
Figure 8. The extent of the Marshes 18,000 years before the present
   day (BP), 5,000 years BP, and currently.
Figure 9. Men fishing in a *shakhtura*.
Figure 10. Jassim paddling a *guffa* (2013).
Figure 11. Tahseen Ali Kadhim al-Asadi, fourteen years old.
Figure 12. Abu Haider, guide and friend of Jassim's.
Figure 13. A traditional *mudhif*.
Figure 14. Women returning to their village with boats laden with reeds.
Figure 15. Wesal, 1978.
Figure 16. A *mudhif* under construction.
Figure 17. Inside a modern *mudhif*.
Figure 18. Satellite image of Marsh extent, 1973.
Figure 19. Satellite image of Marsh extent, 2001.
Figure 20. Buffalo returning home late in the day.
Figure 21. Marsh girl carrying dried reeds.
Figure 22. Young girl and buffalo.
Figure 23. Traditional fishing in Umm al-Ni'aj.
Figure 24. Satellite image of Marsh extent, 2006.

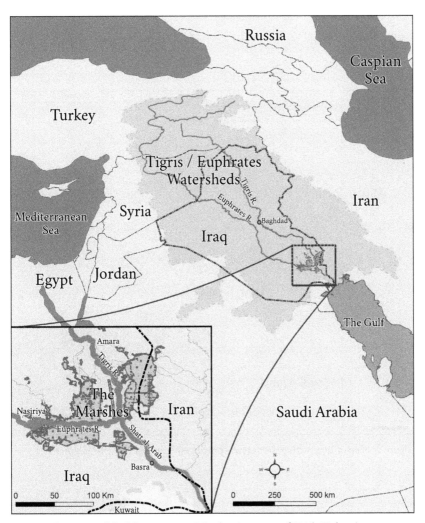

Figure 1. Location of the Mesopotamian Marshes (courtesy of Keith Holmes).

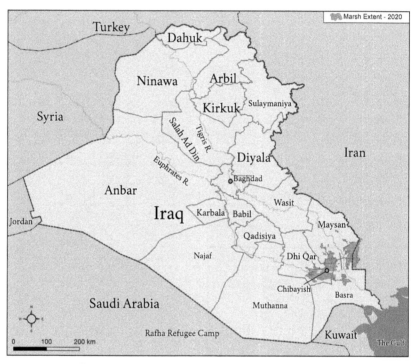

Figure 2. Map of Iraq, with governorates (provinces) (courtesy of Keith Holmes).

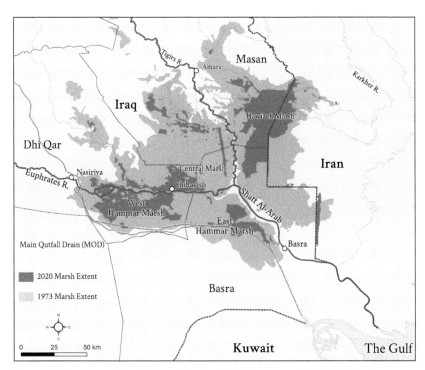

Figure 3. The extent of the Marshes, 1973 and 2020 (courtesy of Keith Holmes).

Figure 4. West Hammar Marsh (courtesy of Mootaz Sami).

It's not often that you see a place that reminds you of nowhere else on earth. The Marshes are like that. They seem to transcend time, not just in the usual way of a glimpse of life as it was decades or centuries ago but it's more as if everything else falls away out on the water with the birds and the sunlight and the sound of the boat.

Jane Arraf, The New York Times Baghdad Bureau Chief, 2021[1]

INTRODUCTION

The common perception of the Middle East by those living in the Western Hemisphere is a land of deserts, oil, and Islam. And yet, in a part of the Middle East once known as Mesopotamia, there are two great rivers: the Tigris and the Euphrates. These two rivers wind their way south from southern Turkey through Syria and Iraq, rather like unruly neighboring strands of a man's long beard, until they meet and form the Shatt al-Arab River in southern Iraq, which then flows through the city of Basra and out into the Gulf.[1] Just north of their confluence, the rivers would often flood their banks to form a large area of connected wetlands in Iraq and Iran known as the Mesopotamian Marshlands (figures 1, 2, and 3).

The Mesopotamian Marshlands were once among the largest wetlands in the world, covering an area of more than 10,500 square kilometers (km^2), roughly the size of Lebanon and larger than twenty-seven countries in the world. During times of extreme floods, the wetlands could extend to 20,000 km^2. They supported a diverse range of flora and fauna and housed a human population estimated between 500,000 and 750,000 by the mid-twentieth century. The region is part of the Fertile Crescent, a semi-circular cultural and ecological land bridge that ranges from Egypt to Syria to the Gulf. The southernmost section of the Fertile Crescent is also known as the cradle of civilization, where agriculture flourished, and modern culture began. A place where 6,000 years ago, writing, mathematics, metallurgy, and hydraulic engineering were invented, and city-states formed. The most important natural resource that fueled the growth and development of southern Mesopotamia was not oil, but water.

The portion of the wetlands lying within Iraq is often termed the Iraqi Marshes. The people living there are mostly Arabs, and the dominant religious group is Shi'a Islam. For centuries, people moved back and forth through the wetlands unfettered by any notion of national boundaries, and many residents continue to have strong ties to Iran.

Biblical scholars consider the Marshes to be the site of the Garden of Eden. The great cities of Ur, where Abraham was born, and Uruk, the largest city in the world in 3200 BCE, were on the Euphrates River and the edge of the Marshes. The two most important religious centers for Shi'a Muslims, Najaf and Karbala, are close by. Indeed, it would be difficult to overstate the cultural, historical, and ecological uniqueness of the region.

The people of the Marshes lived in huts made of reeds. Reed stalks framed the structures, and woven reeds formed their roofs and walls, along with the mats people sat and slept on. Their livelihoods were based on fishing, hunting, and farming. Later, water buffalo were introduced, most likely from India. Marsh dwellers who tend to water buffalo are known as *Ma'dan*.

British explorer Wilfred Thesiger was one of the first travelers from the West to spend significant time in the Marshes. After living there on and off for seven years in the 1950s, he documented his experiences in *The Marsh Arabs*, published in 1964.[2] The account of his exploits and stories about the people of the Marshes are familiar to a generation of geography students in the United Kingdom. Nevertheless, the Marshes remained remote, virtually inaccessible to outsiders, and all but forgotten by the West. The Iran–Iraq War of the 1980s and the subsequent Gulf War and Shi'a Uprising in the early 1990s, all of which took place in and around the Marshes, further limited access to the area.

Thesiger was followed by friend and colleague Gavin Young, who spent time there in the 1960s and early 1970s. In his work *Return to the Marshes* (with photographs supplied by Nik Wheeler), Young provides an in-depth perspective on the Marshes and the people living there.[3] In the last few paragraphs of the book, Young predicts a change coming to the Marshes and to the way of life that had remained unchanged for thousands of years.[4] In this, he was prescient.

Thesiger and Young would likely be shocked to see how the region has been transformed in the past half-century. Wars fought in and around the Marshes altered the landscape and displaced thousands. Massive dam projects in Turkey reduced the flow of both the Tigris and Euphrates Rivers and, in turn, the amount of water reaching the Marshes. Large swaths of wetlands have been claimed for oil and agricultural development. The increasing magnitude and frequency of drought have had deleterious ecological and economic consequences. Most significantly, the purposeful draining of large sections of the Marshes by the government of Iraq during the 1990s, ostensibly to reclaim land for agricultural development and promote economic modernization, almost destroyed the Marshes.

In the fall of 2001, a rather innocuous-sounding technical report was published by the UN Environment Programme (UNEP) in Nairobi, Kenya. "The Mesopotamian Marshlands: Demise of an Ecosystem" was authored by Hassan Partow, a UNEP employee based in Geneva.[5] It was a report that shocked the international environmental community. Using satellite imagery, Partow provided visual evidence of the extent of devastation in the marshlands. Although he acknowledged that upstream dams on the Euphrates and Tigris Rivers played a role, his main conclusion was that the purposeful draining of the marshlands by the Iraqi regime in the early 1990s resulted in the almost complete collapse of the ecosystem and the displacement of hundreds of thousands of people. The report concluded that urgent action was needed to protect the remaining wetlands.

The Iraqi government, unhappy with the contents of the report, lobbied UNEP to stop publication. It was to no avail, however, and publication went ahead. The world had its first glimpse into the disaster perpetrated by the Iraqi regime on the Marshes and its people. It stands as one of the greatest environmental and humanitarian catastrophes of the twentieth century.

1

ORGANIZING THE NETHERWORLD

> There is nobody in the Bani Asad tribe who is not an orator or a
> poet or a preacher or a horseman.
>
> Yunis ibn Habib (Persian poet and literary critic, 798 CE)[1]

Jassim Al-Asadi zipped up his light jacket as he stood next to a dilapi-
dated reed structure in the deserted village of Abu Subbat in southern
Iraq. It was late in the afternoon of December 17, 2003 and the air was
unexpectedly cool. The setting sun revealed a land of dusty tan soil, dried
reeds, and a few desiccated palm trees badly in need of water. The soil
underneath was hard, uneven, and fractured. A few small, green bushes,
able to draw water from deep in the ground, popped up like they had
been purposely placed there to offset the dreary landscape. The horizon
appeared as a straight line almost devoid of color, a monochromatic, milky-
gray sky setting off the pale brown soil. It was eerily quiet.

Jassim's gaze fixated on a large earthen dam, some five or six meters
high, rising in the distance. The Iraqi government built the dam twelve
years before to prevent water in the nearby Euphrates River from flowing
into Abu Subbat Marsh, a small dried-up marsh that was once part of the
large wetland ecosystem covering much of southern Iraq. For most of his
life, Jassim was able to stand in this spot and look out over an enormous
expanse of water, green reeds, and hundreds of birds. The land was now
lifeless, save for a few small bushes.

Jassim wasn't alone. Standing next to him were Ali Shaheen, who
worked for the provincial Ministry of Water Resources, and Azzam Alwash,
director of the newly established Nature Iraq, the only environmental

organization in the country. The three engineers contemplated whether it would be possible to knock down the embankment and allow water to once again flow into the marsh. There was some urgency in their efforts since it was not clear the federal government would support such an action. Breaking down the embankment with picks and shovels—tools that were readily available in the nearby town of Chibayish—would take months. The residents had already tried this approach but to no avail; the soil embankment was too wide and too compact for them to make much headway. Digging a hole in the dam required heavy equipment.

Azzam turned to Ali. "Do you have an excavator? One that we can bring in to break down the dam and allow water to flow back into the marsh?"

Ali was wary. The punishment for contravening government regulations under the previous regime was to be sent to jail—or worse. Iraq was now in transition after the U.S.-led invasion earlier in the year, and America and Britain had assumed responsibility for Iraq's government at both federal and provincial levels. With the country verging on chaos and local militia controlling much of southern Iraq, it was unclear who was in charge. The federal government had worries other than debating whether the Marshes should be reflooded. Ali's boss, the provincial minister, was unlikely to agree without a clear signal from Baghdad. It was one thing to have groups of men with pickaxes and pumps spontaneously tearing down smaller embankments, and quite another to use heavy machinery owned by the provincial government for a much larger task. Iraqi citizens were conditioned to be risk-averse after thirty years under Saddam Hussein.

Ali thought for a few moments before informing Azzam that the use of heavy equipment would require approval from Jassim's employer, the Iraqi Ministry of Water Resources. Only then would he agree.

"Forget the ministry," Jassim responded. "This work is in the best interests of both the people and the natural environment. Nobody will know you had a hand in this."

Ali was still skeptical, telling Azzam and Jassim that there was no way to deliver the machine from the main equipment yard in Nasiriya, 100 kilometers to the northwest of Abu Subbat.

At that point, Azzam intervened, offering to rent a flat-bed loader to transport the excavator. Ali, who wanted to see the dried marsh reflooded as much as Azzam and Jassim did, finally agreed. He knew there might be repercussions, but he also understood that this was the will of the people and hoped the local authorities would back him up.

Azzam phoned a truck rental agency and agreed to a price of 500 Iraqi dinars, approximately $150, to transport the machine from Nasiriya to Abu Subbat. Three days later, the excavator arrived, along with a competent operator from the ministry who knew Ali and was willing to accept the risk of upsetting the government.

When the flat-bed loader reached Abu Subbat, more than two dozen young men from Chibayish joined the excavator driver to view the proceedings. Many wore a white or black *dishdasha*, an ankle-length, long-sleeved robe, along with a red-and-white or black-and-white checked *kufiyya* wrapped around the head or neck. The younger boys wore pants and jackets, much like they would almost anywhere in the world. Six men stood out from the crowd: They all wore kufiyyas, but instead of dishdashas, they sported trousers and jackets or heavy shirts, as if they might suddenly be called into action to defend the excavator. In their arms, they carried semi-automatic rifles.

The men with guns were more than mere spectators; they were there to ensure that government employees did not come and try to disrupt the proceedings, as well as to protect both the excavator and its operator. They stayed on site for more than twenty-four hours, guarding the excavator until it was loaded back onto the truck for the trip back to Nasiriya.

The group watched as the excavator, which looked every bit a prehistoric metal creature, moved on its two wide tracks off the loader. It then turned and proceeded ponderously across the road and over the dry, packed soil toward the embankment. Extending from the cab that housed the operator was an expandable orange arm reaching at a forty-five-degree angle into the sky. Attached to the tip with a pivot mechanism was a second expanding orange arm that dropped straight toward the ground. At the base of the second arm sat an enormous bucket with five teeth—or claws—on one side. It was a cumbersome beast, and once it reached the bottom of the embankment, the excavator slowed to a crawl. The machine struggled to climb the dam, belching black smoke from its diesel engine as it made its way to the top.

With its orange arms and body in stark contrast against the gray sky, the cab swiveled on its base until it was parallel to the top of the dam. The operator then expanded the second arm to a length of seven meters, lowering the

4,000-kilogram bucket to the ground. The claws at the end of the bucket dug into the packed soil and proceeded to scrape and drag, eventually scooping out four cubic meters of dirt. The cab then pivoted 180 degrees and the driver deposited the soil on the opposite side of the embankment. The only sounds were the banging and screeching of metal as the excavator unhurriedly went about its task. Onlookers seemed mesmerized by the sight and were rendered speechless, as if holding their breaths.

When the excavator spun around to dig up the second bucket of soil, excited voices from the young men drowned out the noise of the machine. A few of the men wanted a closer look and clambered up to the top of the embankment. From a distance, they appeared like miniature figures standing on either side of the excavator. The machine continued its laborious movements, dredging the earth with its bucket, lifting the dirt free of the embankment, and depositing it in a mound where it could no longer impede water from entering the marsh.

Three hours later, there was a cut in the side of the embankment roughly four meters high and six meters wide. Soon thereafter, water from the canal started to spill over into the dried marsh, and what began as a trickle soon became a steady flow. After twelve years, water was back in Abu Subbat. As the water returned, shouts of joy emerged from the assembled crowd, followed by dancing and singing. Only the sudden sound of gunfire drowned out all the voices. Fortunately, the guns were aimed at the sky—it was, after all, a celebration.

Two months after water reentered Abu Subbat, a vast shallow-water lake covered the area, the surface broken only by small islands. Remnants of reed huts that once housed village residents stood slightly askew on a few of the islands. There were fish, birds, and aquatic plants, including reeds which provide sustenance to animals and building materials for humans.

On April 21, 2004, Saddam Shayal's family was the first to return. His family and ancestors lived in the area for 500 years until they were displaced in 1992 and forced to move to Yusufiya, just south of Baghdad. Twelve years of hardship and misery. Their island, located on the western bank of the lake, was still visible, although their reed house had been badly damaged when the Ba'th Party security forces burnt it down after the family emigrated. In just two days, Saddam Shayal's family, with the help of friends from Chibayish, built a new reed house at the center of the island. Saddam and his family were elated to be back home.

Almost a year after the incident with the excavator, Jassim hosted the minister of water on a tour of the newly flooded marshes around Chibayish. When they came to Abu Subbat, the minister looked at the large cut in the embankment where water entered the marsh. He turned to Jassim and asked who was responsible for breaking down the dam. Although reluctant to involve his friend Ali Shaheen, Jassim thought it best to tell the minister the truth.

"That was a great idea," the minister declared. "We should use the excavator to knock down other embankments as well."

Jassim broke out in a big smile. The reflooding of the Marshes was triggered when a coalition of U.S., British, Australian, and Polish forces invaded Iraq from the south in late March 2003 and toppled the government of Saddam Hussein. In little more than three weeks, all major cities in the country were under the coalition's control. The country quickly descended into chaos, and with the police also in disarray, crime and corruption became the norm. The invasion may have toppled Saddam and the Ba'th regime, but it also unleashed regional, tribal, and religious factions that had been suppressed for nearly three decades.

Anger at both the economic situation and the occupying forces first turned to protest, and then to violence. The government remained disorganized, with some ministries operating and others shut down. Government buildings were badly damaged during the bombing and looting that followed. It was a tumultuous time, and not many waited for the new central government to make pronouncements about the marshes in the south.

On April 10, 2003, a dozen young men brandishing pickaxes and small water pumps demolished dams and embankments that controlled water flow into Abu Zareg, a small, dry marsh that was once part of the much larger Central Marsh located just north of the Euphrates River. They worked quickly, fearing the federal or regional government would interfere in their attempt to return water to the marsh. After a few hours of digging, they engaged the pumps to help push water through a cut in the embankment. Soon it was trickling back into the marsh. Three days later, the channel was complete and a river of water, roughly three meters wide, connected the outer canal and the former wetland. Abu Zareg was, once again, a marsh. In fact, word of their efforts spread, and soon marshes near Basra were also reflooded. The government could do little to stop the spontaneous effort to resuscitate the wetlands.

Still, not all the marshes that existed when Jassim was a young boy could be reflooded. In the 1970s, the government began to drain areas of marsh for the exploration, extraction, and transportation of oil. During the two decades that followed, military road construction, conversion of wetlands to agriculture, and further oil development reduced the extent of the Marshes to roughly 70 percent of their original size in 1970. For some of the remaining wetlands, reflooding took time. Many martyrs who fought against the regime in the early 1990s were buried in the Marshes near Chibayish, the largest town in the region. It was agreed that their remains should be exhumed and transferred to Wadi al-Salam, the large cemetery in the holy city of Najaf, before reflooding could occur. In other cases, as in Abu Subbat, heavy equipment was needed to break down the larger dams.

Some people were not happy to see the water return. For instance, a few wealthy landowners had taken advantage of the lack of water to expand their agricultural holdings, and the last thing they wanted to see was their land reflooded. At a town meeting in the city of Nasiriya one evening, a local sheikh approached Jassim, who was attending as a representative of the Center for the Restoration of the Iraqi Marshes (CRIM), a branch of the Ministry of Water Resources. The sheikh and his tribe controlled a large amount of land south of the town, and past dealings with this sheikh had never been pleasant.

The sheikh peered at Jassim with bulging eyes and pursed lips and made it clear that he would not allow the ministry to reflood any land controlled by his tribe.

"The Marshes are crucial to the town's livelihood," Jassim replied. "How can we convince you to let us reflood this area?"

Jassim knew he had the support of the other six tribes in the region to open the main intake site that would allow water to flow from the Euphrates River south to West Hammar Marsh. Without this main water inlet, the hydrological system would be inadequate to provide enough water to replenish the marsh. Unfortunately, the intake site was on the sheikh's land.

After a few seconds of pondering, the sheikh told Jassim that he would allow the ministry to open the intake site and reflood the marsh only if an embankment was built to protect the remainder of his land. The sheikh was adamant that his tribal lands would not be reflooded without receiving something in return, and he was powerful enough to dictate the terms.

"I promise you I will look into this," Jassim told him. Not wanting to engage any further, Jassim turned and walked away.

The next day, he approached the minister of water resources and convinced him to authorize the building of an embankment around the sheikh's land. Less than a month later, the embankment was finished, upon which Jassim went to see the sheikh and ask for his permission to open the intake site. The sheikh refused.

"But you gave me your word—an embankment in exchange for the land to be reflooded," Jassim pleaded.

The sheikh, however, was unwavering.

Jassim left the meeting feeling angry and dismayed—but not for long. On April 9, 2004, the first anniversary of the fall of the Ba'th regime, young men from four different clans gathered on the banks of the Euphrates River and defied the police. They broke down the soil embankment south of Chibayish with shovels and pumps, and most of West Hammar Marsh was once again filled with water. The police were more bemused than angry and stood idly by. The sheikh still had his dike, but the rest of West Hammar was, once again, a marsh.

The Marshes would not die. Not if Jassim and other marsh dwellers had anything to say about it.

Figure 5. Traditional fishing in the Marshes (courtesy of Mootaz Sami).

Not long before the Euphrates joins the Tigris in southern Iraq to form the Shatt al-Arab River, it passes through the district of Chibayish, part of the governorate of Dhi Qar and one of over a hundred administrative districts in Iraq. In the middle of the twentieth century, Chibayish, which means "many small floating islands," was a rural area dominated by wetlands. There was one main market town in the district, also named Chibayish, with a population of roughly 11,000 people. From above, the town appeared isolated, lonely, and trapped by water.

The south side of town adjoined the Euphrates River. Past the river, as far as the eye could see, lay the Hammar Marsh, a green and blue expanse of wetlands that stretched out forever. On the backside of the town, north of the river, was the Central Marsh. When the wind rose, the reeds and water would moan and murmur and the marshes would become unsettled, as if awakening the 'afrit and the jinn, those mythical demons and supernatural creatures that haunted the waters. Nestled deep within the boundaries of the district were more than sixteen hundred islands built from reeds that humans piled on top of one another, year after year. There was water everywhere. The rivers, canals, marshes, and forests of reeds and sedges were all part of a freshwater ecosystem that was quite rare in the mostly dry and arid Middle East. Little had changed in over 4,000 years. This was an area rich in biodiversity, with a myriad of migratory and resident birds, abundant fish life, unique species of mammals, as well as a few dangerous species, such as wild boar and poisonous snakes. Lion and hyena once populated the district, but no longer, having been ravaged by hunters.

Most people living in Chibayish at that time belonged to the Bani Asad, a semi-nomadic tribe that immigrated to Mesopotamia from the northern Arabian Peninsula around 600 CE. They initially settled in Babylon before moving to Chibayish. Although not a large tribe, they were an eminent one; members of the Bani Asad were among Prophet Muhammad's *sahaba*, or close companions, and one of the Prophet Muhammad's wives—Zaynab bint Jahsh—was a Bani Asad. They are Shi'a Muslims renowned for their linguists and poets, and members of the tribe are known by their family name: Al-Asadi.

In the 1950s, the Bani Asad of the Chibayish District lived much like the Sumerians and Akkadians millennia before: they engaged in traditional fishing methods using small nets, raised water buffalo, and lived a quasi-communal life. Homes were built primarily of reeds, whether in town or in the middle of the adjacent wetlands. When one gazed out over the marshes, there was nothing to block the view, regardless of direction. The horizon could appear vague and indeterminate on cloudy days but mostly emerged from the dark of night as if someone drew a sharp green line in the middle of a canvas of soft gray. Closer in, the blue skies and green reeds reflected off the water, as if two worlds were unfolding in front of one's eyes.

There was only a single dirt road in the entire district, a simple path less than two kilometers long that abutted the Euphrates River in the town of Chibayish. It had a grand-sounding name for a small dirt track: Corniche al-Chibayish. There were no cars or trucks in the entire district; boats were the only means of transportation unless you counted walking up and down the corniche. Summer days were very hot and winters comfortably warm, although winter winds could sometimes blow hard and cold, rendering the marshes dangerous for small boats. Throughout the district, nights evinced a blueish hue as stars that appeared like magical art forms in the sky reflected off the water. There was a peaceful rhythm to life and a harmonious balance between humans and nature.

Fatima Abbas, dressed in a traditional robe and black hijab and just shy of seven months pregnant, paddled her wooden canoe, or *mash-huf*, through the marshes to cut young reeds that could be used as fodder for her three cows. It was July 1, 1957, and the waterway was teeming with villagers and peasant women, poling or paddling their canoes along the channels and through the grasses. When she reached her destination, where the bright green reeds became almost impenetrable, she rolled up her sleeves, grabbed a handful of reeds, dispatched them with a sharp sickle, and placed them inside her mash-huf. After an hour of working, she felt pains in her stomach. Fearful at first, Fatima stepped out of the mash-huf and lay on a bed of reeds; thirty minutes later her son was born. She placed the baby in the mash-huf and paddled back to the middle of town, where she carefully exited the canoe and slowly made her way home, carrying the baby in one arm and green grass for the cows in the other. The boy was born in nature and would forever be part of nature. His name was Jassim, and he would be associated with the Al-Asadi family.

Soon after they can walk, children in the Marshes help with family chores. They stack reeds, feed animals, and become intimately familiar with the pleasures and dangers of the natural environment. Jassim, like many in his community who spent half their lives in or on the water, soon developed an intense emotional relationship with the Marshes that never waned. He was the oldest child in what would become a family of nine: Fatima, her husband Muhammad, four boys, and three girls.

When Jassim turned six years old, he would accompany Fatima every Friday on her trips to gather green reeds for the cows or the yellow reeds used to weave mats. He sat at the front of the boat and loved both paddling and the movement of the boat as his mother directed them through the channels of water, singing the entire time. He watched as birds built their nests, seemingly chirping in harmony with Fatima, and as the buffalo swam across the water toward the grasslands. When Fatima reached her destination, she cinched up her skirt, rolled up her sleeves, and began cutting the reeds. Green frogs hopped in and out of the water, making a sound that, to Jassim's ear, sounded like *cactus, cactus, cactus*. Occasionally, he would spot a turtle walking deliberately through the reeds and then sliding into the water, or a snake slithering through the grasses—he knew well enough to avoid the snakes, for many were poisonous.

The reeds in the Marshes were almost as important as the water to residents. It wasn't only houses that were built from reeds; so, too, were the shelters and fences used to house and constrain animals, and the mats used for sitting and sleeping. Even the school was built from reeds. On the other hand, the younger, more delicate reeds, which the Arabs call *hashish*, were food for many of the animals . . . Unprocessed nature: raw but sustainable.

Jassim began formal schooling at the age of six. Wearing a white dishdasha, he often paddled his small mash-huf—called a *chileakah*—four kilometers each morning to class, where the wooden desks were placed on reeds stacked high enough to withstand spring floods. The students sat in a semi-circle around the teacher, and accompanying them was a complete army of animals that included frogs, snakes, turtles, fish, and birds. The animals and birds frequented the floors, walls, pillars, and ceilings. Their movements never ceased.

When he turned twelve, Jassim was allowed to fetch reeds on his own in his chileakah, with room enough for one or two people. His father made a small sickle for him, and on cutting and returning with his first batch of reeds, his father exclaimed, "Now you are a man, Jassim!"

The trips to collect reeds were hardly a burden. For Jassim, it was more nature tourism than work. He learned from the older men about the geography of the area, the various waterways (which were not always easy to see from the bottom of a small canoe), and the hidden passageways between the Euphrates River and the Central Marsh. It was a joyful time for a boy.

During school term, Jassim and his younger brother Hazim lived with their grandmother Massouda on a farm just outside Chibayish, on the edge of the Euphrates and the Hammar Marsh. Their reed hut in the Marshes was too far away for a daily commute to school. The main crop on the farm was rice, supplemented by corn, watermelon, cowpeas, and assorted vegetables. Like every farm, there were a few cows and chickens to provide milk and eggs. Once a year, two of his relatives would load up a large wooden boat with reed mats woven by family members. They would then pole the boat toward the northwest, along what was known as Hassan's Brook, to the Beni Hassan area of southern Nasiriya, where they would exchange the mats for young rice seedlings to plant on the farm. When the rice was harvested, usually in mid-November, the family would transport part of the crop to their residence deep in the Marshes, on an island they shared with eight other families.

Jassim grew up in the Marshes, and they became part of his heart and soul. The farm, on the other hand, helped develop his sense of justice and equity. With its roots in land reform initiated by the Ottomans during the middle of the nineteenth century, a quasi-feudal system was firmly entrenched in southern Iraq. In the Land Law of 1858, the Ottomans instituted a land tenure system in Mesopotamia that made it consistent with the rest of its empire and ensured more revenue for the sultan in Istanbul. The state continued to own all land, although individuals could be granted title deeds that would allow them rights of ownership. They, in turn, would be responsible for tilling the land and paying rent to the government. Collective ownership of land, however, was not allowed; customary land given to families in accordance with tribal tradition became private land.

The result of the Land Law was that sheikhs, as tribal leaders, registered their tribal lands and obtained title to them. In some cases, tribal groups refused to abide by the new law, and powerful individuals from outside the region obtained the titles. With the fall of the Ottoman Empire after World War I, these title holders became official owners of the properties. The farmers essentially became serfs to the sheikhs, and they were required by law to perform certain duties—mainly to earn money for someone else. There was no way out of this indebtedness for the farmers; they were subject to arrest if they stopped farming and left the land. The system was a throwback to the Middle Ages, and it changed the power structure of southern Iraq, particularly in the tribal areas.

Farmers in Chibayish, like Jassim's grandmother and her family, had to pay rent and a portion of their revenues to both the tribal leader and to a large property management group. Jassim's grandparents were given the property according to tribal division, and documents show they paid taxes to the government on agricultural yields from 1927 onward. They had settled on the land long before, but the entrenched legal system carried over from the Ottoman Empire meant they did not have legal title to the land. The inequities inherent in this centuries-old system created a peasant class of farmers and had a significant influence on Jassim's view of the world.

Jassim's grandmother would often recite poems to him about the brutality of the feudal lords and the battles the men in the al-'Awwad[2] clan fought with the tyrants of the feudal system. Every Thursday at school during the flag-raising ceremony, the teacher read a poem by Arab poet al-Akhtal al-Saghir, encouraging students to defend the land against feudalism. Through these experiences, the young Jassim developed a heightened sense of inequality and injustice.

Nevertheless, in addition to the feudal landlords imposing their will on the farmers, there was another political issue affecting the residents of the Marshes: the Ba'th Party was gaining national attention and would soon take over the country. It was a party that promoted Arab nationalism and viewed the Shi'a of southern Iraq, with their ties to Iran, as an existential threat. But political maneuvering in Baghdad was hardly a major concern to a young boy living in the Marshes.

Jassim split his time between his grandmother's farm and his father and uncle's reed house in the Central Marsh. He looked much like all boys of his age—medium height, slender build, and a shock of black hair—with one exception. His brown eyes evidenced an intensity that left one with the impression he was not only serious about life but had a depth of understanding far beyond his years. His father Muhammad ran a general store in town that sold food, animal feed, fishing and hunting supplies, and other assorted goods. Many of his customers were themselves small merchants, filling their boats and poling them deep into the Marshes to sell to the Ma'dan. The economy was primarily based on farming, fishing, buffalo breeding, and rug weaving, all of which provided a sustainable lifestyle—one with much freedom, despite having to obey the feudal lords, who demanded both a percentage of sales and rent.

Jassim enjoyed learning and reading. Books provided a welcome break from the tedium of family work and gave him a broader sense of himself and his place in the world. In middle school, he developed a fondness for poetry and literature. He was, after all, a Bani Asad, and his grandmother Massouda had been a poet. The library became his refuge, and he immersed himself in stories, real and imaginary. Gibran Khalil Gibran, Mustafa Lutfi al-Manfaluti, Badr Shakir al-Sayyab . . . Jassim loved them all. He cherished the writers, artists, and poets almost as much as the Marshes themselves. The artist inside was emerging, and soon drawing and carving in clay became his daily routine.

Jassim's first poem was penned at the age of twelve. The teacher was so impressed that she asked him to read it in front of the entire school. Jassim stood on a small stage at the front of a room filled with students, adults, and, of course, small animals. A few community leaders were present, including a representative from the Ba'th Party, which had recently taken control of the government of Iraq. Jassim received an ovation at the end of his presentation, along with a gift from the Ba'th Party. It would have surprised nearly everyone to learn that the poem was the young boy's initial foray into political protest and a veiled criticism of the Ba'thists.

I will not die, hey cowards, I will not die
as long as in my broken nose there is a wind
as long as in my wounded heart there are drops of blood
as long as my people wake up from silence
from sleeping on stones,
I will not die
even if they put chains on my hands
and they threw my body to the monsters and to the dogs
and they arbitrarily killed me, like a hawk with a crow
but I will not die[3]

Two years later, one of his former teachers saw him on the street, pulled him aside, and said, "Jassim, you never should have written that poem."

Fortunately, the Ba'thists failed to comprehend his true intentions. Once his protests and writings moved further afield, however, they did notice, and the repercussions were swift and severe.

Figure 6. Jassim, 17 years old (courtesy of Jassim Al-Asadi).

Figure 7. Marsh village, circa 1975 (courtesy of Nik Wheeler / Getty and Canadian Iraqi Marshlands Initiative, CIMI).

In those ancient days, when the good destinies had been decreed, and
after An and Enlil had set up the divine rules of heaven and earth,
then . . . the lord of broad wisdom, Enki, the master of destinies, . . .
founded dwelling places; he took in his hand waters to encourage and
create good seed; he laid out side by side the Tigris and Euphrates,
and caused them to bring water from the mountains; he scoured out
the smaller streams, and positioned the other watercourses.

Debate between Bird and Fish (Sumerian Disputations)[4]

The *Enuma Elish* is a Babylonian creation myth written on seven stone
tablets almost 4,000 years ago. The tablets were discovered in 1849 in present-day Mosul in northern Iraq during an excavation of the Royal Library of
Ashurbanipal, the last great king of the Assyrian Empire. In the story, Enki, a
name that means "Lord of the Earth," is the Sumerian god of wisdom, fresh
water, intelligence, trickery and mischief, crafts, magic exorcism, healing,
creation, virility, fertility, and art. It is a portfolio that any present-day parlia-
mentarian would be proud of. He is depicted on stone carvings as a man with
a long beard, wearing a robe and a horned cap. Most notably, two streams of
water run from his shoulders, representing the Tigris and Euphrates Rivers,
which are said to be formed from Enki's semen.

As Lord of the Earth, Enki has a prominent place in Sumerian creation myths, and as the god of fresh water, he is invariably linked to the
Tigris and Euphrates Rivers. In the Sumerian poem, "The Debate Between
Bird and Fish," fresh water did not exist until Enki created the Tigris and
Euphrates in the mountains and caused water to flow. The poem—one
of seven Sumerian Disputations—also depicts other gods influencing the
rivers. Enlil, the god of wind, air, earth, and storms, was upset with the
noise humans were making on earth and took his vengeance by causing
floods in the lower part of the river basins when he copulated with the hills
of the earth. It is unclear whether the poem refers only to the great flood
that destroyed civilization (a common theme in many creation myths), or
to seasonal floods, or both. In any case, humans are still making noise, and
spring floods continue to occur in the lower part of the Tigris and Euphrates, tempered today by multiple dams on the upper sections of the rivers.
And while floods may wreak havoc on villages and crops and change the
course of rivers, they are also vital in replenishing and cleansing marshes
further downstream.

Almost every culture in the world posits one or more creation myths that explain the beginning of the world and how order arose from chaos. The *Enuma Elish* is particularly important since it may have been the inspiration for the Book of Genesis from the Christian Old Testament and the Hebrew Bible. The parallels are many: creation occurred over six days with the seventh day being a period of rest, there were great floods, and the Tigris and Euphrates appear as central characters in both accounts. Nevertheless, the story of the origin of the Tigris and Euphrates is much more straightforward in the Judeo–Christian tradition: God created the Garden of Eden along with a main river to water the garden; this river had four branches, and two of these were the Tigris and the Euphrates (Genesis 2:14). Scholars believe the Garden of Eden was in southern Mesopotamia, between the Tigris and Euphrates Rivers. Their argument is strengthened because the lands mentioned in the Book of Genesis as being close to the Garden of Eden lie east of the original city of Babylon, in what are now known as the Iraqi Marshes.

Ten thousand years ago, rain and snow that fell in the Taurus Mountains in eastern Turkey would often find its way into three budding rivers: the Kara Su (also known as the Western Euphrates), the Murat Su (the Eastern Euphrates), and the Tigris. The Kara Su and the Murat Su meander through the mountains of Turkey for roughly 450 and 700 kilometers, respectively, before meeting to form the Euphrates, which then flows at a leisurely pace through southern Turkey, Syria, and Iraq (figure 1). Once the river reaches southern Turkey, it looks from above like a single, spindly tree branch, lacking any offshoots save for a few places where the river has been diverted to form a reservoir or a canal for irrigation. There are multiple dams on the river, but only one, Haditha Dam, is in Iraq.

Haditha Dam is one of the largest in the Middle East, producing 30 percent of Iraq's hydroelectricity. Located 280 km northwest of Baghdad, the dam was one of the first sites that U.S. and coalition forces seized during their invasion of Iraq in 2003 in order to prevent its destruction. South of Haditha, the Euphrates flows past the important Shi'a religious centers of Karbala and Najaf and then encounters a system of barrages, levees, and canals built over thousands of years. The result is a seemingly random pattern of waterways as the river meanders, breaks apart, and then merges again, aiding both irrigation and flood control. It is an ill-behaved beast. Between the third and second millennia BCE, the Sumerians used this to their benefit as they dug channels and canals, taking advantage of how the river would break off from its original course and form a new channel—a process known as avulsion. More importantly, while the Euphrates' unruly behavior might not have allowed for a deep, navigable channel, the spillover from the river helped form the largest freshwater wetlands in all of western Asia.

After passing Nasiriya and tipping its hat to the ancient city of Ur, the Euphrates veers east. At this point, water flow can be quite variable due to annual weather cycles, upstream withdrawals, and return flows from fields of rice that abut the river. Indeed, during dry years, the depth of the river might be less than one meter. More importantly, the water is too polluted from agricultural runoff and municipal waste discharge for human consumption. The Euphrates then meanders past the town of Chibayish and eventually merges with the main branch of the Tigris River at Qurna to form the Shatt al-Arab River, which flows south through Basra and out into the Gulf. The waters of the Euphrates are still important to southern Iraq, particularly for irrigation. During wet years, its waters overflow and replenish the Marshes. However, it is no longer the grandiloquent master of great city-states such as Babylon, Uruk, and Ur. It staggers to the end of its journey in poor health, having relinquished its power to upstream dams and divergences while receiving in return only the by-products of great human endeavor.

The Tigris River begins on the slopes of Mt. Ararat, not far from the Murat Su. It then flows south, roughly parallel to the Euphrates. For a short distance, it forms the border between Syria and Turkey before entering Iraq. From above, the Tigris begins looking like a tree with its branches chopped off one side—no channels leave or join its western bank. Soon after entering Iraq, the main river flows through the Mosul Dam, just north of the city of Mosul. Despite being much smaller in size than the Haditha Dam, it is equally as strategic and produces over a third of the country's hydroelectricity.

Once it departs Mosul, the Tigris wriggles and writhes, squiggles and squirms, seemingly never able to quite decide what course to take. Tributaries from the east add to its flow as it makes its trip south to Baghdad, where it appears as if a magnetic attraction is pulling the main channel of the Tigris toward the Euphrates, only to repel it a few kilometers later. As the river nears the city of Amara, the Tigris looks as if it has grown propagation roots: streams flow east and west into what were once vast wetlands, and the main channel of the Tigris continues to Qurna, where it meets the Euphrates.

The Tigris and Euphrates Rivers are not the longest rivers in the world—that distinction is reserved for the Nile River—nor do they form magnificent gorges like the Yangtze in China or the Colorado in the United States. Their combined discharge into the Gulf is relatively small, at less than 0.5 percent of the mighty Amazon River. In terms of the history of civilization, however, they are without peers. Mesopotamia is an Ancient Greek word that means "the land between two rivers." Those rivers are the Tigris and Euphrates, where civilization began—or at least one of the places where civilization began. The Sumerians were a non-Semitic people who migrated from West Asia or North Africa and settled in the northern valleys of the Tigris and the Euphrates. By 3500 BCE, they had moved south, to what is now southern Iraq. They developed city-states, and their largest city, Uruk, had an estimated population of between 40,000 and 80,000. They also invented writing, metallurgy, mathematics, hydraulic engineering, the plow, textile mills, and mass-produced bricks and pottery. The Sumerians have the oldest known recorded law, the Code of Ur-Nammu, written around 2100 BCE. Scholars now believe that civilization arose at roughly the same time in different regions of the world. These other regions may come close to equaling Mesopotamia in their overall contributions, but none surpass it. And water was the resource that made it all possible.

The slope of the land in the far south of Iraq is almost imperceptible, the elevation barely above sea level. From Baghdad to the Gulf—over 600 kilometers—the elevation decreases by a mere thirty-four meters. The deposition and subsequent build-up of sediments allowed the Euphrates to transform itself from a single channel to a series of small channels and, eventually, to freshwater marshes. The Tigris exhibits the same behavior but also manages to maintain a distinct channel that was navigable until the middle of the last century. The result was a large area of relatively shallow wetlands and a highly productive ecosystem, due to the excess nutrients. It is a region highly vulnerable to the vicissitudes of dry and wet years, desiccation, and flooding—a region that is now called the Iraqi Marshes. In dry years, there were separate wetlands. In wet years, these were connected into one marsh system.

The temperature in the Marshes can be blisteringly hot. Weather in Chibayish, a large town that borders the Euphrates and sits between two major marshes, reached 56 degrees Celsius (133 degrees Fahrenheit) in the summer of 2020, one of the hottest temperatures ever recorded on Earth. The high temperatures are accompanied by high rates of evapotranspiration, resulting in salty water and salty soils; water evaporates and the salt is left behind. In some cases, shallow marshes dry out completely. The elegant natural solution to this problem is flooding.

In the northern part of the basin, wet years brought large volumes of water rushing down the rivers that cleansed the Marshes of salts. They also provided enough water for smaller marshes to connect physically and ecologically. Once human development began along the rivers and in the Marshes, the floods became problematic. They destroyed structures, damaged crops, and killed people. Upstream dams built over the last seventy years have stabilized the water flow in the rivers, reducing the vulnerability of human systems. But these same dams have had an enormous impact on the ecology of the Marshes. The lack of pulses of water raging down the rivers results in an increased buildup of salts, negatively affecting the productivity of both plants and animals. It also means that the various marshes are no longer connected, even in wet years, and they can no longer be considered part of one marsh system. Rather, they survive as separate marshes and separate ecosystems.

Jassim was eleven years old in 1968 when the Ba'th Party came to power in Iraq. He soon learned that membership in the Party came with certain privileges. The opposite was also true. Political considerations affected all aspects of life regardless of age or ability. Kheri Hussain was a school friend of Jassim's who would often visit relatives in Basra, the second-largest city in Iraq after Baghdad. Kheri and his family were members of the Ba'th Party, although, for most young teenagers, party affiliation was not an abiding concern.

During one of his visits to Basra, Kheri's uncle taught him how to play chess. When Kheri returned to Chibayish, he taught Jassim and a few others how to play the game. Aside from Jassim, most were fourteen or fifteen years old. The boys carved out chess pieces from chalk and wood, and then made a playing board out of a large piece of white cardboard, using black dye to distinguish the squares. They would get together at different houses after school, three or four days a week, and play well into the evening.

Jassim soon became quite adept at the game and by the time he was sixteen felt confident enough to enter a local tournament sponsored by the Sports Committee for the town of Chibayish.

Jassim progressed through the first few rounds of competition with ease and made it to the finals. His opponent was Wathaah al-Asadi, the son of the mayor of Chibayish. The mayor and his family were also staunch members of the Ba'th Party. Muhammad Hanoon, the head of the Sports Committee for Chibayish, presided over the final match and awarded the prizes: trophies for the finalists, one large and one small.

Jassim handily defeated Wathaah in the finals. When the trophies were handed out, Jassim received the small one and Wathaah was presented with the large one. Jassim was confused and walked over to Muhammad.

"Why did I receive the smaller trophy when I won the match?" he asked.

Muhammad simply replied that Wathaah was the mayor's son.

Jassim was more confused than devastated. After all, he had won the tournament.

Three years later, he had the opportunity to speak with Muhammad again. "I still don't understand why I was given the smaller trophy," he said to the sports director.

Muhammad appeared sympathetic but reiterated that Wathaah's father was not only mayor of Chibayish, but also a senior member of the Ba'th Party, which meant he wielded a lot of power.

"I know you won," Muhammad admitted, "but I thought it best for everyone to give Wathaah the larger trophy."

It wasn't an explanation that satisfied Jassim. However, he did understand. The inequities between members of the Ba'th Party and everyone else were becoming very clear.

Kheri Hussein left Chibayish to attend the University of Basra and study chemistry in 1974, the same year Jassim moved to Baghdad to pursue his engineering degree. Although he was still a member of the Ba'th Party, Kheri surreptitiously joined Hizb al-Da'wa, a popular Islamist party that was critical of the Iraqi regime. This was not uncommon. Jassim knew others who publicly were members of the Ba'th Party but privately supported al-Da'wa. He would often see a group of engineering students meeting under a large tree on his university campus along with people he knew were al-Da'wa members. Many Shi'a Muslims moved their allegiances to al-Da'wa in the 1970s while maintaining their membership in the Ba'th Party. They could not publicly acknowledge this, since the risks were too great, but privately there was little question about where their loyalties lay. By the late 1970s, members of Hizb al-Da'wa were being arrested for supporting an insurgency against Saddam. In 1980, membership in al-Da'wa was banned by the Iraqi regime, and all former and present party members were sentenced to death.

Kheri excelled in his studies and continued playing chess. However, he was unable to keep his membership in al-Da'wa secret for long. Shortly before he graduated from university in 1979, he was arrested and executed.

The story of the formation of the expansive wetlands in southeastern Iraq relates to the story in the *Enuma Elish* involving Marduk, the god of fresh water, and Tiamat, the goddess of salt water. Just as there was a battle between the two gods in the creation of the Earth, there has long been a battle, literally and figuratively, between salt water and fresh water for control of the Marshes.

The Pleistocene Epoch, or what is commonly referred to as the Ice Age, came to an end roughly 14,000 years ago. Large mammals, such as the enormous woolly mammoth and the saber-toothed tiger, were almost extinct. Humans had migrated from Africa to Eurasia and Southeast Asia and were starting to enter North America. Glaciers still covered much of the Northern Hemisphere. Sea levels were 130 meters below their present level—the height of a forty-story building—and the global average temperature was eleven degrees Celsius lower than it is today. It was a cold climate, with much of the Earth's water locked up in ice. There were no wetlands in southern Mesopotamia, only an arid landscape of grasses fed by a few lonely streams. Then, however, a major shift in climate occurred: Earth surface temperatures began to rise, glaciers melted, sea levels rose, and rivers became swollen and raging. The Gulf, which had previously been nothing but a dry basin, began to flood and sea levels continued to rise—sometimes quite dramatically. Within the relatively short geological time period of 10,000 years, the Gulf not only flooded but extended all the way to the ancient Sumerian city of Ur, very near the present city of Nasiriya and some 300 kilometers north of its present shoreline (figure 8).

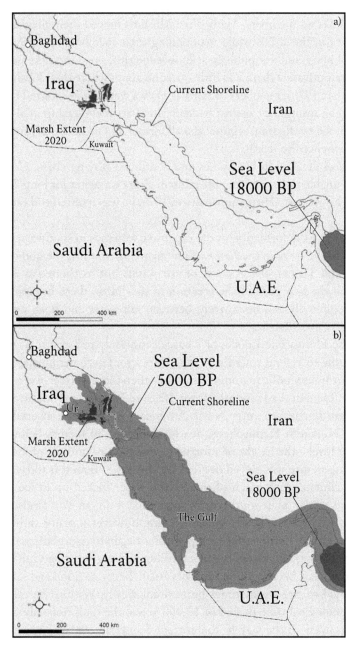

Figure 8. The extent of the Marshes 18,000 years before the present day (BP), 5,000 years BP, and currently (courtesy of Keith Holmes).

The city of Ur is important to the history of the Marshes. The birthplace of Abraham—and his home until he left for Canaan—was once a major center for trade and agriculture. Boasting two ports, one on the Euphrates River and one on the Gulf, it had access to fresh water for drinking and irrigation, and sea trade with regions as distant as India. The city therefore became a hub for culture, agriculture, and construction, and at its peak had a population of 60,000. However, this is no longer the case. Today, the Euphrates is sixteen kilometers to the east, the Gulf is 300 kilometers to the south, and the landscape is dry and dusty. The only historical structures that remain are the Ziggurat of Ur—once a large community center—and a few walls that were built to keep out marauding tribes.

"The Lament for Sumer and Ur" is a poem purportedly written by Nanna, the moon goddess and patroness of Ur. It is one of five dirges inscribed on clay tablets by the patron gods or goddesses of five different Mesopotamian cities. The dirges were discovered in 1918 in the ancient city of Nippur, north of Ur and in the Marshes. The tablet inscribed with "The Lament for Sumer and Ur" is now in the Louvre in Paris. It tells the story of how the gods of heaven (An), water (Enlil), and storms (Enki), and the goddess of the mountains (Ninhursag) destroyed the city and sealed Ur's fate. The resulting drought shifted the centers of power in Mesopotamia and changed what was a saltwater marsh to a freshwater one, with the Gulf receding to its present location.

When the drought ended, fresh water rushed down from the north and pushed out the remaining salt water to form the present-day Marshes. However, the goddess of salt water Tiamat was nothing if not a forceful deity; she left behind enough salt to cause problems. Her legacy continued long after "the North wind bore her away into secret places," as written in the Lament.[5]

Being thus disconnected from the ebb and flow of popular culture—whether 4,000 years ago or today—the Ma'dan and other marsh dwellers have their own way of life, their own set of beliefs, and their own stories.

Under the roof of their reed house, lying in front of a stove burning dry sticks and dried buffalo dung, Jassim closed his eyes and listened as his grandmother patted his shoulder and sang a mournful and affectionate hymn, reassuring the young boy that his enemies were sick and unable to harm him. Their enemies lived in a distant land, in the cities and the wastelands; the angels would protect the family and the villagers.

"Ours is a world of creation," she sang. "It is known in our hearts."

The hymn was about allaying a young boy's fear of the unknown and providing a sense of comfort in the unique and isolated environment of which they were a part. Theirs was a community of people who lived mostly separated from the world, where distant lands were only a myth. Water, green grasses, and endless sky all contrasted with the idea of "the Other"—the outside world. The Maʿdan sense of identity is influenced by their stories of how the Marshes were formed, how the Maʿdan came to be on this earth, and what lies beyond the Marshes.

One such story is from *al-rawat*,[6] told through generations of Marsh dwellers. It relates that one day, God removed all the darkness and created the heavens and the earth. On the same day, he made an emerald-green jewel—several times larger than the heavens and earth—that sparkled in the light. When God cast his gaze upon the jewel, it became water. As He looked in awe at the water, it began to boil and foam, giving off smoke and steam. The smoke created the sky, and the land was born from the foam.

The earth at that point was nothing more than a single dish. God rolled it up and made it into seven earths, and from the heavens, he sent down an angel who held the earths high over his head, with one of his hands clutching the east and the other clutching the Maghreb.[7] The angel held up the seven earths with his body and weighed them to make certain they were equal. God then sent down from Paradise a bull with 70,000 horns and 40,000 feet who proceeded to stand on the angel's back. But all the bull's feet could not fit on the back of the angel, so God created a green sapphire the size of the seven earths and rested the bull's feet on top of the stone. The sapphire stone needed a home and companions, and so God then created Nona, the great whale, whose name is Lootia. God placed the stone on Lootia's back, and moving through the sea, the whale created the wind. Afterward, God transformed the sapphire into the Marshes.

The story of the sapphire stone includes elements of obedience to God, wisdom and goodness, the seven regions of the world, and land, water, and fish. It is no wonder the residents of the Marshes believe the place they inhabit emanated from a sacred historical spirit with tinges of magic and energy. The evidence is revealed in front of them. A palate of clouds, sky, grasses, and water.

The lives of those living in the Marshes are inextricably linked to the water. Their enemies aren't human—at least most of the time. They come in the form of floods and drought. And, on occasion, in the form of ghosts.

Figure 9. Men fishing in a *shakhtura* (courtesy of Mootaz Sami).

> One day, a shadow sneaked up behind me and hit my head violently, leaving me badly injured. I no longer walk straight and feel like the ghost still haunts my body.
>
> Haider al-Hatemi (former gravedigger), 2019[8]

Located in the holy city of Najaf, Wadi al-Salam, or "the Valley of Peace," is the largest cemetery in the world. It covers over six square kilometers and contains millions of bodies. Najaf is also the burial place of Ali ibn Abi Talib, Prophet Muhammad's cousin and son-in-law and the first imam of the Shi'a sect. For this reason, Shi'as from across the globe want to be buried at the venerable site. Many who work in Wadi al-Salam believe the cemetery also houses ghosts, called *tantal*s. Some, like Haider al-Hatemi, even claim to have been attacked by them.

There are many supernatural creatures in Arabic and Islamic folklore. Many inhabit the underworld and have an amorphous character: they can be evil in one context and good in another. At one extreme are the *shaytan*s who are evil spirits that try to lead humans astray; they are invisible devils who manifest evil. Jinn are more ambiguous. They are generally considered to be malevolent ghosts but can also, on occasion, be benevolent intermediary spirits, like the genie in *One Thousand and One Nights*. The jinn seem to occupy a middle space between humans and angels/demons. *'Afarit* are demons that have been formed from smoke and fire. They live in a subterranean world and, while not fundamentally evil, are often considered spirits of the dead. Some believe they are an evil type of jinn, and the Qur'an usually refers to them as 'afarit of the jinn.

Tantals, on the other hand, are different from both. Although Haider believed it was a tantal that bashed him over the head, these ghosts generally act as protectors of sacred places rather than evil spirits. Nevertheless, it can be a mistake to ignore tantals, as Haider discovered. There are occasions when a tantal acts more like a practical joker. In Samir Naqqash's novel, *Tantal*, a child becomes obsessed with seeing and touching the tantal. He wants to know for certain that it is real. He dreams about the tantal and speaks with others who have seen it. One day, walking along the seashore as the light of day is fading, he meets a giant. The giant becomes playful and tries to entice him into the water, but the boy runs away. When he returns home, he realizes the giant was a tantal. His dreams unfold before him, and he understands that the tantal is something both real and not. It is a moment of being that lasts a lifetime.[9]

The roots of the character al-tantal date from the same period as the Babylonian creation myth *Enuma Elish*, in which Enlil, the god of wind, air, earth, and storms, unleashes the great flood in southern Mesopotamia in response to humans becoming too noisy. The Old Testament version says that God flooded the earth because human beings were corrupt. Regardless of the cause, the consequence was a great flood that devastated towns and cities in southern Mesopotamia.

The flood myth told by the Ma'dan is slightly different from better-known versions. According to the Ma'dan, in ancient times there were three large kingdoms in southern Mesopotamia: al-'Aker, Abu Shathar, and Hafeez. They were built on large islands *(eshan)* that rose high above the wetlands in the north of the Central Marsh. The kingdoms had beautiful architecture, with curved, ornate temples adorned with gold and jewels. Strong fences, or *masnayat*, were built around the cities to protect them from floods. These were ringed by palm groves and fruit trees. It was God's paradise on earth, and life flourished there. When the people became disobedient, corrupt, and noisy, God grew angry and caused a massive earthquake that destroyed the kingdoms. This was followed by a flood to rid the world of all humans, save for a few who were righteous. As the human population increased, God felt it was important to protect the ancient cities and their treasures and sent down fairies and tantals to guard them. There are many sites in the Marshes known as al-eshan, or "the islands", many of which contain pottery, glassware, coins, gold pieces, and statues that date to the Sumerians, Akkadians, and others. Tantals watch over them all, but only one is guarded by Tantal Hafeez. Tantal Hafeez lives on Eshan Hafeez, an island sometimes known as Hufaidh, protecting the gold and relics. On some nights, fires can be seen appearing as distant lights emanating from Eshan Hafeez. But no one approaches. Not anymore. A group of Ma'dan set out to explore the source of the light one night, and all were temporarily blinded as they approached the blazing fire. Their limbs stiffened and they were fortunate to escape with their lives. Another resident—so the story goes—admitted to exploring Eshan Hafeez but claims that when he tried to leave with some relics, he was unable to find his way home until he returned the pieces. In fact, those who have visited Eshan Hafeez and survived say they are unable to find it again.

Other possible explanations exist for the fires on Eshan Hafeez. When reeds and aquatic plants decompose, they release methane gas, which can ignite and rise up in the sky. Moreover, some Ma'dan set fire to dry reeds in winter to allow for young, green grass to grow, which is then eaten by buffalo and cows. Nevertheless, no one can explain definitively what causes the lights on Eshan Hafeez. And no one dares to find out.

The Ma'dan often build their reed houses on al-eshan, since the higher ground offers some protection from the periodic extreme floods that can destroy villages. The support structures for the reed houses are large arches of woven reeds called *al-shibab*. The reeds are tied together with ropes, also made of reeds. To glorify God in the *mudhif*, there is always an odd number of al-shibab—as few as five or as many as twenty-seven. Ancient artifacts are often discovered when digging the holes for al-shibab. But no one dares build on Eshan Hafeez.

Confronting a tantal anywhere in the Marshes can be risky. They have been known to cause deformities, disabilities, and insanity to those who encounter them. But they have also developed friendships with the Ma'dan and their buffalo. Unsalted grilled rice bread *(hinaya)* is often given as a peace offering to the tantals by the Ma'dan. And although tantals are not afraid of much, they dislike needles, stitching, and any machine made of iron. Moreover, once its face is exposed to humans, a tantal quickly recedes into the Marshes.

Jassim's mother constantly warned him about tantals when he was a young teenager. During the day, they are rarely a problem. At night, it is a different story. Jassim often ventured out on Friday nights to meet his friends at a local coffee shop and watch Iraqi wrestler Adnan Al-Kaissie (who was known as Billy White Wolf on the professional wrestling circuit in North America), on one of the few black-and-white televisions in town. To get there, he had to walk across the river and patches of marsh—known hiding places for tantals. Before crossing the water, he would remove his clothes and place them on his head to keep them dry. Once he reached the other side, he would put them back on. The trip always frightened him. Memories of those nights still send a chill down his spine. Still, Jassim was fortunate. For although he would hear their moans on occasion as he waded through the marshes, the tantals remained hidden.

2

THE COMING STORM:
WATER AND POLITICS

It is said that before entering the sea
a river trembles with fear.

She looks back at the path she has travelled,
from the peaks of the mountains,
the long winding road crossing forests and villages.

And in front of her,
she sees an ocean so vast,
that to enter
there seems nothing more than to disappear forever.

But there is no other way.
The river cannot go back.

Gibran Khalil Gibran[1]

It was late summer 2013 and Jassim, clad in blue jeans and a light blue, short-sleeved shirt, was standing in a round basket boat with a flat bottom, less than two meters in diameter, called a *guffa*. The basket was constructed of woven reeds tied together with palm fiber rope, the bottom waterproofed with bitumen, and the sides, which were only a half meter in height, covered with a thin layer of dried mud. Holding a long paddle, he was trying to row the guffa down the Tigris River. As he began paddling, the basket spun around, first one way and then the other. Eventually, he figured out how to direct the guffa in a straight line—at least until the current in the river became stronger, in which case he spun around again. Guffas date back to Assyrian times (800 BCE) at least and were used to ferry people and goods across rivers throughout the region until the middle of the twentieth century. Larger ones could be three meters in diameter and a meter deep and could carry up to thirty people. Until the 1970s, guffas ferried people across the Tigris River in Baghdad.

Following close behind Jassim was another strange vessel called a *kalak*—a traditional raft made of strong reeds or wood that stays afloat by attaching inflated goat skins to the sides. It floats downriver with the current and, because of a very low draft, is useful in shallow rivers and streams. Depending on the size and the number of goatskins, a kalak can hold up to seven tons of animals and goods. Kalaks were often seen on the Tigris between Mosul and Baghdad and on stretches of the Euphrates River north of Fallujah where the currents were relatively strong. They were not, however, vessels for going upstream or across rivers. Once the goods reached Baghdad or Basra, the kalak was dismantled and the operator had to find other means to return to his point of origin. In the late 1880s, a trip from the Turkish border to Basra along a rapidly flowing river might take only nine days. The trip home was much longer.

The third member of this group of odd vessels is the *tarada*, an elegant canoe with long tapered ends that reach toward the sky. Traditionally made from acacia or mulberry wood coated with bitumen, these were long boats—up to thirteen meters—and could carry as many as twelve persons. Regal in appearance, taradas were used as war canoes and were also the boats of the sheikhs in the Marshes. Today, however, powerful sheikhs find automobiles preferable to a tarada, and the boats are used only for ceremonial purposes.

These three vessels, along with a flotilla of modern inflatable boats and kayaks with their passengers, embarked on a month-long, twelve-hundred-kilometer journey, traversing the Tigris River from Hasankeyf, Turkey, to Qurna, Iraq, where the Tigris River meets the Euphrates. Hasankeyf is an ancient settlement and rock fortress almost four thousand years old that was built above the river. In 1981, the Turkish government declared Hasankeyf a natural conservation area. In 2008, the World Monuments Fund listed Hasankeyf as one of the most endangered archeological sites in the world. But that hardly mattered because in the summer of 2020 the town was completely submerged by the reservoir behind the newly constructed Ilisu Dam.

The flotilla was organized and coordinated by Nature Iraq to develop awareness in local communities about threats to the Tigris River, and to encourage regional partnerships to protect water resources. The group hosted presentations, photo exhibits, clean-up events, and community-based art projects along their route. A major focus was the impending construction of the Ilisu Dam, a project that would not only flood the ancient city of Hasankeyf, but displace thousands of people, endanger biodiversity, and significantly reduce the amount of water flowing downstream into Iraq.

During his time at the Iraqi Ministry of Water Resources, Jassim, realizing that there were constraints on what he could accomplish as a government employee, worked closely with Nature Iraq. In 2011, he retired from the ministry and began working at Nature Iraq full-time. While the organization advocated for environmental issues in general, most of its work focused on helping preserve the Marshes of southern Iraq. At Nature Iraq, Jassim had the freedom to combine his technical expertise as an irrigation engineer with his love of the Marshes to help people whose lives depended on a continuous flow of water into the wetlands. By this time, the Marshes were in trouble.

Figure 10. Jassim paddling a *guffa* (2013) (courtesy of Nature Iraq).

The vast wetlands in southern Mesopotamia of Jassim's youth were shaped thousands of years before by geological and meteorological forces on Earth, possibly with the help of gods and goddesses like Enki and Tiamat. The most obvious features were ubiquitous water and dense forests of reeds. There were sections of permanent marsh that contained at least some water throughout the year, fringes of marsh with wet soils that were generally covered by vegetation, and seasonal marsh that was dry during part of the year. Land used for agricultural purposes, particularly for rice production, might have been flooded at times as well, and appeared from above as a marsh.

The extensive carpets of reeds and bulrush that stretched out endlessly toward the horizon around Chibayish concealed a vast array of birds, fish, and other wildlife. Resident birds included flamingoes, egret, heron, kingfisher, and gulls. The bird population swelled in winter, as large flocks of geese, ducks, and grebe flew in from the north. Small birds delicately hopped on water lilies with white and purple flowers, blue kingfishers hovered a few meters above the water, and birds of prey or flocks of ducks could be seen high above, highlighted against the blue sky. The Marshes were a treasure trove of water and food for the masses. Almost 300 different species of birds were identified in the 1970s, including endangered and vulnerable species such as the White-headed Duck, the Basra Reed Warbler, and the Marbled Duck.

Higher ground in the Marshes and the muddy banks on the border of the wetlands also housed a range of mammals. Aside from the domesticated water buffalo, wild boar was one of the most recognized species. These dangerous beasts could cause significant damage to crops, although they were mostly hunted out by the late 1970s. Lions were also present in the Marshes, at least until the end of the nineteenth century. They, too, were hunted to extinction. More prevalent were the bandicoot rat and the Mesopotamian gerbil, both endemic species. Otter and mongooses were abundant during the middle of the twentieth century, as were lizards, skinks, turtles, and snakes. The biodiversity was impressive, particularly for a region that was surrounded by desert.

The largest open water lake was Umm al-Ni'aj in the northeast, near the Iranian border. Some sections of the lake were six to seven meters deep and had the longest and sturdiest reeds in all the Marshes. These reeds were used to construct arches for the mudhifs - the largest houses that acted as guest houses, meeting places, and ceremonial centers. The deeper water also harbored submerged aquatic vegetation, like eelgrass and hornwort, along with fish species such as bin, mullet, and various types of carp, some of which had been introduced from elsewhere. Before 1990, 60 percent of the fish consumed in Iraq came from the Marshes.

Figure 3 shows the extent of the Marshes in 1973, based on one of the first sets of satellite images of the region. It shows three large, distinct marshes and a few smaller ones. The three large marsh areas are divided by the Tigris and Euphrates Rivers. They include Hammar Marsh, south of the Euphrates; Hawizeh Marsh, east of the Tigris; and the Central Marsh, nestled between the rivers, northwest of the confluence of the two rivers at Qurna.

"Hammar" means "redness," which refers not to the water but the color of the surrounding soil. The Marsh appeared as an elongated stretch of wetland, as if being squeezed between the desert to the south and the Euphrates River to the north. Fed by the Euphrates, it stretched over 100 kilometers from just south of Nasiriya almost to Basra. It had sections of deeper water as well, where fish were abundant.

Hawizeh Marsh lay east of the Tigris River and straddled the border between Iran and Iraq. Hawizeh is an ancient term meaning "God's possession." The origin of the name is unknown, although there is a village with a similar name in Iran. The northern portion of the Hawizeh contains Umm al-Ni'aj. Hawizeh was the most remote and pristine of the three marshes, although it was later marred by war during the 1980s, and its southern section has been the site of substantial oil development in the past four decades.

Hawizeh's wetlands lakes were replenished from multiple branches of the Tigris River and Iran's Karkheh River. Having multiple sources from two river systems reduced the vulnerability of the marsh to upstream control—at least that was the case until 2008, when Iran built a dike along the border, posing a major threat to the long-term survival of the marsh.

The Hawizeh Marsh was drained by closing water feeders and outlets such as the al-Kasareh River. This process ensures that water flows through the system, preventing too much salt buildup. As a result, Hawizeh exhibited the highest water quality of any marsh, and was home to the largest number of fish species and migratory birds.

The third large marsh was the Central Marsh. Its original name was the Marshes of Qurna. The Central Marsh nested within the arms of the lower Tigris River to the east and the Euphrates River to the south. Residents described its shape as a bride with her arms raised (signifying the Tigris and Euphrates), her head turned to the north, and her wide dress spayed out toward the south, the fringe just touching the Euphrates River as it turned to the east into the Marshes. The major cities of al-Islah, al-Maymonah, and Qurna acted as a triangle delimiting the marsh boundary. Like Hawizeh, the Central Marsh also included a series of smaller marshes, such as Chibayish, Odeh, and al-Kabirah, and occupied a central position between Hammar Marsh and Hawizeh. The Central Marsh was also characterized by numerous villages of buffalo breeders. Unlike people living in villages on the edge of the Marsh with a fixed location, the buffalo breeders were semi-nomadic, moving from place to place depending on pasture conditions and water quality.

Most notable among the smaller marshes was Abu Zareg, or Prosperity Marsh, which was once part of the Central Marsh and entirely dependent on water from the Tigris River. Abu Zareg, shaped a bit like Aladdin's genie emerging from his lamp in *One Thousand and One Nights*, was home to many buffalo breeders. The marsh is surrounded by agricultural land, which in recent years has led to many conflicts between the buffalo breeders and farmers over access to water. It was the first marsh to be reflooded after the fall of Saddam Hussein in 2003.

The Ba'th Party and the Iraqi Communist Party worked hard to expand their bases of support in the 1970s, particularly with the Shi'a in southern Iraq. Recruitment focused on attracting youth members. A religious alternative to the Ba'th Party, however, was Hizb al-Da'wa, a popular Islamist party with strong ties to Iran. Muhammad Baqir al-Sadr helped found al-Da'wa and was its intellectual theorist. He criticized the Ba'th Party for their ideological secularism on the one hand, and extreme policies in dealing with Shiites on the other. When Saddam expelled thousands of Shi'a to Iran in the late 1970s for no other reason than their ethnicity, even though their ancestors had inhabited this geographical area for hundreds of years, al-Sadr turned his wrath on Saddam as well.

Tahseen Ali Kadhim al-Asadi was a school friend of Jassim's from Chibayish. Ali, Tahseen's father, was an educator and director of a primary school in town. The family was well known for both its integrity and religious convictions. Tahseen was fascinated with photography, and Jassim would often accompany him when he took photos of the Marshes, the animals, and the Ma'dan. They became great friends, swimming in the canals, learning to play chess, and exploring the Marshes. Devoutly religious, Tahseen took political beliefs one step further and became a member of Hizb al-Da'wa in 1974, at the age of sixteen. Becoming a member of an Islamist political party in Iraq at that time put family and friends in jeopardy, so it was little surprise that Tahseen kept his activities secret—even from Jassim.

The rapid growth of al-Da'wa in southern Iraq during the 1970s represented a threat to the Ba'th Party. Shi'a clerics such as Baqir al-Sadr were a public source of dissent against the regime, particularly since they disdained Arab nationalism. The conflict with al-Da'wa wasn't simply due to a clash of ideologies, although this was enough reason for the Ba'thists. Al-Da'wa had widespread support in the south and could mobilize thousands of its members for mass protests against the government. It also had strong ties to Iran, which was worrisome as Iraq prepared for war. The Ba'thists started targeting al-Da'wa members in 1972, sending them to prison—or worse. The conflict reached a climax in 1979, when al-Da'wa, with support from Iran, mounted an uprising against the Ba'th Party. This led to the eventual arrest and execution of Muhammad Baqir al-Sadr in April 1980. Kheri Hussein, Jassim's friend who taught him how to play chess, was another person who paid the price for being a member of al-Da'wa.

Jassim approached Tahseen's father in school one afternoon in late 1974. He had not seen Tahseen for over a week and wondered whether he was ill. Ali explained that his son had decided to move to Basra to obtain better schooling; it was a spontaneous decision and Tahseen didn't have time to tell his classmates he was leaving. He was now living with relatives in the city. Jassim was mildly surprised that his friend left Chibayish without saying goodbye, but he remained close to the family and served in the army with Tahseen's younger brother, Ihsan. Jassim took the school director at his word and had no reason to expect Tahseen had joined al-Da'wa. Despite their earlier close friendship, Jassim never saw Tahseen again.

In 1980, shortly before he completed his engineering degree, Jassim was strolling past a news shop in Baghdad when he glanced at the official Iraqi newspaper, *al-Jumhuriya*. The date was April 28, and Jassim was surprised to see two photos of Tahseen on the front page under the byline, "A Member of the Puppet al-Da'wa Gang Talks to the Republic About

Methods of Deception and Misinformation." The article described Tahseen's upbringing in Chibayish, his love of the Marshes, and his eventual move to Basra to work for Hizb al-Da'wa. Tahseen's role had been to travel back and forth from Basra to Kuwait, promoting al-Da'wa and meeting with prospective donors. The risks were considerable, more so once membership in the party of Hizb al-Da'wa became illegal.

Tahseen was arrested in early 1980 while on a trip to the holy city of Najaf in central Iraq. Jassim later learned from the family that although Tahseen had been interviewed by a reporter for one of the regime's newspapers, the story was fabricated to be critical of al-Da'wa. The article ended by describing Tahseen's short stay in prison and his eventual execution. Jassim's knees buckled when he read the story. He knew from his own experience how Tahseen had been treated, and that he had likely died under torture.

Tahseen's parents, on the other hand, had no idea their son worked for al-Da'wa. After the outbreak of the Iran–Iraq War in September 1980, the Iraqi government carried out harsh security measures against families whose children had Islamic affiliations, and Tahseen's family was among them. His father was transferred out of the education ministry into a lower-level job in the Ministry of Transportation. Tahseen's brother Ihsan, who was then in the army and worked as an engineering supervisor at the air force base in Nasiriya with Jassim, was arrested and thrown in jail for two years for criticizing the regime and not divulging his brother was a member of al-Da'wa. Ihsan spent time in five different jails, three of them in Baghdad. During his case's investigation period, he was placed in solitary confinement and tortured. Once the court case was over, he was sent to a jail in Nasiriya to serve the remaining year of his term. He was put in a room with four others and allowed only one hour of exercise per day— although the torture at least had stopped. He was also allowed one visitor per week. His incarceration hardened Ihsan's hatred of the Ba'th Party, and a decade later he would join the Shi'a revolt against Saddam.

Ihsan managed to survive Saddam's jails and returned to Chibayish to work as an engineer before fleeing the country with his family in 1991. In early 2021, he returned to Iraq and took Jassim to the old Kadhim al-Asadi family home in the Marshes. Jassim told Ihsan stories of the games he and Tahseen used to play and about their explorations of the Marshes. Over four decades later, he still mourns his friend.

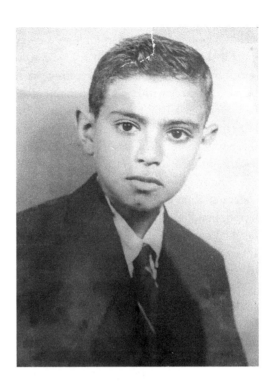

Figure 11. Tahseen Ali Kadhim al-Asadi, fourteen years old (courtesy of Ihsan Ali Kadhim Al-Asadi).

One of Jassim's fondest memories of his youth, besides the freedom he felt paddling through the Marshes, was the day he met Mahasin. Jassim was fifteen years old and out collecting reeds in the Marshes on a cold winter day. His feet were freezing, his teeth chattering, and his joints stiff. His mother had begged him not to go out with the boat that day because it was bitterly cold, but although the cows didn't need more reeds, Jassim was not deterred. He went out in his chileakah whenever he could.

The green reeds were barely above the waterline, so he had to wade through the cold water and reach well under the surface to cut the plants. As he paddled through the narrow waterway in search of more reeds, he found himself in a small lake, edged by a carpet of young, bright green plants. Their reflection off the water made the lake look bright green as well. There was only one other chileakah on the lake, and it was empty. As he looked around, he saw a girl about his age harvesting the soft green reeds while wading through the water.

As he paddled closer, he called out, "Good morning, can I share this place with you to harvest the *'angar*?"

"Of course," she replied. "There is enough for both of us."

She looked vaguely familiar. Jassim rowed toward her, gently hopped out of his canoe, and began harvesting the reeds, all the while exchanging periodic glances with his new acquaintance. Her name was Mahasin. She had thick, black hair, plump lips, and wide, soft eyes. She lived on an island not far from his home, but they had never met face to face. Mahasin was three years older than Jassim, and offered to steady his boat while he collected and loaded the reeds.

The method of harvesting the 'angar is to cut the plant under the water and then leave it floating on the surface. Once cut, the reeds are collected, bundled, and lifted inside the boat. After helping Jassim, Mahasin finished cutting and soon her boat was piled high with reeds. Jassim was amazed at how quickly she worked, wading through the water in a dress that was cinched around her waist.

Before departing, Mahasin turned to Jassim and asked whether they could meet again the next day.

With a very heavy heart, Jassim said, "No, I won't be able to come until next Friday."

Mahasin promised to be at the same spot the following week. Jassim was enthralled. He had found another reason to spend time in the Marshes, and it wasn't long before he wrote Mahasin a love poem. One Friday, while they were alone, he read it to her:

Let me narrate from your cheeks,
and from the delight of life from your wrists
and from the breasts that perch on your chest, I see it,
through the dress that pushes me to you,
and from the mouth like a purple flower,
I am the bee that lives in you,
and from hair as dark as night seems
like a wave coming down to your side.
I am the captain craving your beautiful harbor and beach,
let me play the days as a melody
and I will drink wine from your eyes
and dance like a little bird,
as he flirts with his mother from above in the tree branch.[2]

Jassim kept up his friendship with Mahasin until he left for university three years later. By that time, the political environment in Iraq had changed, affecting Jassim and many of his friends in the Marshes.

Indeed, not everyone found the Marshes an enchanted place. Jassim's brother Hazem, five years younger, was quite content to stay on their grandmother's farm when he had the choice. He also preferred the activities in their father's store to those of the birds and the buffalo in the wetlands. Although Hazem and Jassim spent much time together as children, exploring the Marshes was something Hazem engaged in only reluctantly. During the times when Hazem would join his mother in the Marshes to cut and collect reeds, he left the paddling to her. His only interest was the small, very tender young reeds that he could chew on. Otherwise, he wanted firm ground under his feet.

When Jassim was twelve years old, he and Hazem decided to journey into the Marshes to cut down reeds for their mother. Cutting the reeds was not a problem, but placing the heavy load inside a narrow mash-huf was a different matter. It was all about balance. When their boat was loaded, they clambered aboard and started paddling. On their first attempt, the boat flipped over, dumping both the boys and the reeds back into the water. There wasn't any danger, since both boys could swim and the depth of the water was only one meter, but both were despondent as they watched their pile of reeds float away. On the other hand, the mash-huf, while still afloat, was full of water. Neither boy knew how to remove the water and continue their journey.

A passer-by stopped to help. He showed the boys how to flip the boat over to remove some of the water, hold it steady, and then, with one person at each end, move the boat quickly back and forth to allow the water to splash out—at least enough so that they could paddle themselves home. Before returning, the boys practiced the procedure a few times, knowing it would likely happen again. Hazem was now even more convinced that his place was on land.

Hazem followed in Jassim's footsteps and attended the University of Technology in Baghdad. When he finished his degree in mechanical engineering in 1983, he joined the army, much as his brother did. The Iran–Iraq War was still underway, but he was more valuable as an engineer than a soldier on the frontlines. He moved from Chibayish to Baghdad and eventually to Basra, where he served in the navy. After he was discharged in 1986, Hazem decided to move back to Baghdad. His first job was as a cashier at one of the fancy hotels. Soon after, their father Muhammad loaned him some money so he could set up a carpentry business with a friend. Hazem then sold his share and became co-owner of a small food market. He eventually sold this as well and made enough money to open a shop that sold spare parts for heavy machinery— mostly Caterpillar, a company that builds engines for everything from farming to mining. He was a good businessman, and the shop remains a thriving business today.

These are brothers with the same upbringing, who went to the same university and served in the army, and yet pursued very different paths. One was happiest in the middle of the Marshes, and the other much preferred being in the city, on dry land, surrounded by heavy equipment.

To say that Iraq has had a tumultuous history since independence would be a considerable understatement. The country was almost condemned to be this way when France and the United Kingdom signed a secret agreement partitioning the Arab states of the Ottoman Empire even before the end of World War I. For almost 400 years, the Ottoman Turks controlled a large swath of land around the southern and eastern Mediterranean and Black seas. When they sided with Germany during the war and then were defeated, their empire was divided among the conquering powers. France, Britain, the United States, and, in lesser roles, Italy and Russia, met at the Paris Peace Conference in 1919 and negotiated agreements on control and oversight of Ottoman lands, which included all of Mesopotamia. The British focused on access to the one resource that could change nations: oil. They believed it existed in the northern part of what was known as Greater Syria, near Mosul, and the southern part, near Basra.

The British proposed to combine three provinces of the Ottoman Empire—Mosul, Baghdad, and Basra—into one Kingdom of Iraq under British rule. The French could have Lebanon and the area north of Mosul (what is now Syria). The United States raised some objections, but they were late to the party and had fewer imperialist ambitions. President Woodrow Wilson's principles of democracy and self-determination notwithstanding, the British moved ahead with their mandate. Kurdish and Armenian sympathies at the conference had fallen by the wayside. The British were interested in Kurdish areas that were within their proposed boundaries for Iraq since such areas not only contained oil, but also provided a buffer zone to USSR interests in the northeast. The general feeling was that the Kurds would welcome British rule—a rather cavalier miscalculation. In the end, the new Kingdom of Iraq would be divided into three distinct ethnoreligious regions: a non-Arab, Kurdish region in the north; a Sunni Muslim region in the heart of the country, including the capital Baghdad; and a Shi'a Muslim region in the south. It was a recipe for disaster.

After the war, Iraq was administered under British control with King Faisal I installed as the monarch. Sunni Muslims were a minority in Iraq but were given the most senior administrative posts and effective control of the federal government. The Arabs had been promised self-rule by both the British and French, but both powers were still thinking about expanding their empires, not shrinking them. In 1920, peaceful protests against Britain began in Baghdad and led to an all-out armed revolt in the south. These were followed by an appeal from Sheikh Mahdi al-Khalissi, the supreme Shi'a leader in the country, to boycott elections that might legitimize British rule. Although the British were able to suppress the revolt, the elections were a failure. At the same time, the Kurds in the north of Iraq revolted against the British. The new diverse and divided country worked together for the first time, at least temporarily, as a foil against British rule. This forced the British to negotiate the Anglo–Iraqi Treaty of 1922, which allowed for local self-government but gave Britain control of foreign and military affairs. The Kingdom of Iraq gained independence in 1932, although the UK retained military control well into the 1950s. Oil was simply too important.

The British hold on Iraq was a tenuous one, and their desire to create a system like the British Raj in India was dependent on support from not only the leaders of the three very disparate ethnoreligious regions, but from the wealthy landowners as well. This included many tribal sheikhs who wielded considerable power by virtue of the Ottoman Land Law. They were a force to be reckoned with, particularly in the south. Indeed, by the late 1950s, less than 1 percent of the population of Iraq owned 55 percent of the cultivated land, while 64 percent of the rural population owned less than 4 percent. Jassim's family suffered under this inequitable system, and it is no surprise that he was affected by it. It was a system that begged for change.

The Marshes were not immune to the vagaries of land laws. Flooding from the Tigris and Euphrates deposited rich soil that was ideal for agriculture—provided there was adequate water. There were vast agricultural areas adjacent to the Marshes, which translated into a continued demand from wealthy families to drain existing marshland and convert it to agriculture. While overall agricultural productivity in Iraq increased in the four decades after independence, the system worked against peasant farmers, many of whom were forced to abandon the land and move to cities such as Baghdad and Basra in search of economic opportunities. Not surprisingly, the inequities in land allocation and overall economic welfare spawned social and political movements that influenced many in Iraq and, in particular, a young boy growing up in the Marshes.

In 1958, only a year after Jassim's birth, King Faisal II was deposed in a military coup and Brigadier General Abd al-Karim Qasim assumed the reins of power in Baghdad. The king, the crown prince, and other members of the royal family were captured, taken to the palace courtyard, and killed. The monarchy was finished, but this did not make Qasim's task an easy one. Kurdish residents in the north were clamoring for independence; the Shi'a in the south were concerned about increasing marginalization and the pan-Arab[3] focus of the new government; and union workers and peasant farmers, supported by the Iraqi Communist Party, or ICP, were pressuring the government to address the issues of inequality in Iraqi society. The attraction of the ICP for many was less about its Marxist–Leninist ideology than it was about the crucial need for social reform in Iraq. The ICP was the oldest political party in the country, and it would soon be the largest—a party with many members, but little power.

Communist Party representation at high levels in the military or government was non-existent. Still, Qasim needed the support of the ICP and, for a time, there was an unstable peace between the two. Added to the mix of vocal groups demanding more reform were small factions within the military who had their own ideas of how best to rule the country. In the end, Qasim turned against the reform-minded organizations and institutions, outlawing most political parties, and punishing those who criticized his regime.

President Qasim was torn between two worlds. On the one hand, he desired social reform and even used oil revenues to assist in poverty reduction; he promoted education and tripled the number of students enrolled at all levels; he increased the number of hospitals in the country; and he enacted worker protection laws and raised the minimum wage. However, on the other hand, Qasim was anathema to powerful interests in the south. He also misunderstood what impact the fight for Kurdish autonomy in the north would have on the central government in Baghdad. Most importantly, he overestimated the amount of support he had in the military establishment. Qasim knew that there were rumblings of dissent and that a coup attempt might occur, but he mistakenly believed that his troops would be able to control the situation. This miscalculation cost him his life.

At the time of Qasim's accession to power, the Arab Socialist Ba'th Party was still a fledgling political party that had a small—albeit effective—leadership role in the military. One of the main figures was Brigadier Ahmad Hassan al-Bakr, a strong Arab nationalist with considerable contempt for communism. One day of fighting and it was over: Qasim was brought before a military tribunal, convicted, and shot.

The Ba'th Party originated as a strong Arab nationalist and Arab socialist party in Syria in the late 1940s. By 1954, it was the second largest party in that country. The Iraqi Regional Branch of the party was formed in 1951 and comprised mainly Shi'a Muslims who had been disenfranchised under British rule and believed in greater social equality; these individuals slowly migrated to the Iraqi Communist Party. The Ba'th Party in Iraq became Sunni-dominated and staunchly pan-Arabist. Despite having a small number of members, many were well-placed in the military. They were disquieted with Qasim's package of social reforms and outraged at his relationship with the communists, who were opposed to Arab nationalism.

A year into Qasim's presidency, the Ba'thists made their first coup attempt. One of the young team members was Saddam Hussein. Qasim was injured, but survived, and the coup failed. Ba'th Party leaders, including Saddam, fled to Syria. The die had been cast, however, and it was not long before the Ba'thists increased their numbers, consolidated power, and tried again. In 1968, they succeeded. They soon found, however, that toppling a government was not the same as running one.

The Ba'th Party were initially content to share power with other Arab nationalist groups. Abd al-Salam 'Arif, head of the Iraqi Revolutionary Command Council (the equivalent of a cabinet in a parliamentary democracy), was not a Ba'thist but had broad support throughout the military. The council elected him the second president of Iraq, and his vice president was Hassan al-Bakr. Competing interests made for a slightly disjointed attempt at governance, but there was one thing they both agreed on: the need to eliminate Qasim's supporters, who included members of the communist party. Over the next year, three thousand were killed. And it was only the beginning.

President 'Arif was killed in a helicopter crash in the spring of 1966. For the sake of continuity, the council elected his brother, 'Abd al-Rahman 'Arif, to replace him. It didn't matter; al-Bakr and the Ba'th Party were now in control of the military and had wide support with the public. Abdul Rahman was a weak leader and easy to manipulate, which is exactly what the Ba'thists wanted. In 1968, the Arab Socialist Ba'th Party took over the country in a bloodless coup. Al-Bakr, now the head of the Ba'thist Revolutionary Command Council, became the fourth president of the Republic of Iraq. The Ba'th Party saw initial advantages to working with the Iraqi Communist Party. They desired closer ties to the Soviet Union, which the ICP could provide, and would benefit from the broader international network of the ICP. The Communist Party, in turn, felt that the Ba'thist socialist ideology was close to their own, and that this would ultimately be more important than the Arab nationalism of the previous administration. They were, however, mistaken.

Under al-Bakr, the deputy chairman of the revolutionary command council and vice-president of Iraq was Saddam Hussein. By 1975, Saddam was head of security and had de facto control of the government. It was clear by this time that the Ba'thists were ardent promoters of Arab nationalism,

Figure 12. Abu Haider, guide and friend of Jassim's (courtesy of Mootaz Sami).

an ideology that alienated both the Kurds in the north and the Shi'a in the south. They also demanded fierce loyalty to the Ba'th Party. "Unity, Freedom, Socialism" might have been the motto of the party, but oppression became its guiding principle. Oil revenues allowed for rapid economic growth and modernization, but this was accompanied by severe repression of critics, both outside and within the party. With more oil revenues, the need for a link to the Soviet Union as a supplier of military weapons was less important: the Ba'thists could buy their arms anywhere. This did not bode well for the Iraqi Communist Party. For while the Ba'th Party was rooted in socialist idealism and intellectualism, it soon became anti-Western, anti-intellectual, and anti-communist.

There were of course those Iraqis who desired a more democratic form of government, who treasured the art, writing, and poetry of their ancestors, and who were painfully aware of the inequities inherent in the semi-feudal agricultural system. People who also believed in a non-sectarian political system and were willing to speak openly about it. These individuals naturally gravitated to the Iraqi Communist Party (ICP). Jassim was one of them, and in 1973 he joined the Iraqi Democratic Youth Federation, the youth wing of the ICP. Although only sixteen at the time, Jassim made a secret decision to join the ICP, and none of his family were party members. The Ba'th Party held no attraction for him—the party's strong belief in Arab and Iraqi nationalism too reminiscent of the demagogue who ruled Germany thirty years before. He also saw first-hand the inequities in the system: students in high school who were active in the Ba'th Party were given preference when it came to college admissions, and he found this an affront to his sense of equity and social justice.

Despite his politics, Jassim had long dreamt about moving to Baghdad to attend university. It was the center of Iraqi culture, the seat of political power, and had the best universities in the country. At the end of secondary school, he applied for admittance to the University of Technology, the top engineering school in Iraq. He was accepted and moved to Baghdad in the fall of 1974 to study machine engineering.

Jassim loved the city, although he missed the Marshes and returned there frequently. In addition to attending classes, he visited art galleries and frequented poetry readings. Lodging and tuition were paid for by the government as part of the Ba'th Party program to modernize Iraq. Jassim also received a monthly stipend of 15 dinar (roughly $50), which was supplemented by a similar amount from his family. The additional money allowed him to attend the theater on occasion or to eat at a restaurant.

During his second and third years, he lived in a student dormitory. In his final year, he lived with a friend and his family in al-Thawra, a neighborhood in East Baghdad, which is now called al-Sadr City.

Jassim also became more active in the Communist Party, working as a journalist for the party newspaper, *Tariq al-Sha'b* (Path of the People). He had two roles at the newspaper: the first was to write and produce a section on student activities, and the second was to help with distribution. Iraqi universities were the centers for many student activities and programs related to poetry, painting, music, and sports. They were also the site of demonstrations and protests against the government and in solidarity with other peoples. Every Tuesday, an entire page of *Tariq al-Sha'b* was devoted to student activities throughout the country. As a journalist, Jassim never used his real name when writing for *Tariq al-Sha'b*. Sometimes, articles and editorials were anonymous. Other times, he used a pseudonym: Safwan.[4]

Baghdad was changing. Jassim's personal ideals and goals were based on a strong sense of family and community that was manifest in the Marshes, along with a heightened sense of democratic principles based on the inequities inherent in Iraqi society. These were at odds with the Ba'th Party, which was controlled by Saddam Hussein. In the beginning of 1978, the already strained relationship between the Ba'th Party and the Iraqi Communist Party began to unravel. Iraqi security services started harassing ICP members, and arrests and beatings were common. The situation deteriorated rapidly, and by August many senior party members had been arrested and executed. Others fled to South Yemen, the only communist state in the Arab world, or went into exile in Europe. Naziha al-Dulaimi, the former minister of municipalities, was among them. Remaining ICP members went underground, but it was too late. Saddam's security services had infiltrated the party two years before and they had a long list of members and their activities. The list included Safwan.

On August 26, 1978, Jassim was on a break from university and staying with his family in Chibayish. That morning, he was walking along the Corniche al-Chibayish dressed in a blue dishdasha, heading toward his father's store, when three security men grabbed him from behind, bound his hands, and pushed him along the road toward the regional security center. Jassim knew the three from high school, where they used to take pleasure in harassing other students.

After entering the center, Jassim was taken to a room and shoved roughly into a chair. Leaning on a metal desk directly in front of him was Captain Ibrahim, a large, rugged-looking man with a thick, black moustache and meaty hands, his olive-green uniform stained and somewhat disheveled. Ibrahim was not from Chibayish but was a Ba'th Party loyalist who had been brought in from Mosul to oversee security and ensure no favoritism would be shown to the local population.

The captain raised his head, looked at Jassim, and spat in his face. He then slapped him. Once, twice. In a gravelly voice, he told Jassim that he was a despicable communist and had only one choice: to join the Ba'th Party. Ibrahim picked up a single sheet of paper from his desk. The document stated that Jassim agreed to become a member of the Ba'th Party and that he would never do partisan work again, except for the Ba'thists.

And then the captain said, "If you fail to sign, you will be executed."

He stared at Jassim for a moment before resuming. "Sign it and you go free."

Jassim shook his head and quietly said, "No."

"You won't sign?" Ibrahim asked.

"No." Louder this time.

Ibrahim looked up and nodded to the guards. They grabbed Jassim from behind, wrapped a blindfold around his head, and yanked him to his feet. He was pulled and shoved out of the room, down the hallway, and into a cell where they removed his shoes and socks. He stood for a moment, unsure what to expect.

Jassim heard the whoosh of the strap before he felt the sharp pain in his foot. Again, and again, it slashed his feet, as if he were being bitten by a snake. By multiple snakes. A moment of silence, his feet on fire. And then the kicking commenced—initially to his legs and groin. When he fell to the floor, they kicked his chest and back. He doubled up in a fetal position, with his hands trying to protect his face. And then nothingness.

A few hours later, Jassim woke up in a government hospital. Not for a long stay—just to check for any permanent injuries. The next day, he was taken back to the center and the process was repeated until he again passed out. This time, he was left in the cell.

The torture and abuse continued every two to three days, along with periodic visits to the hospital. Afterward, the demand was always the same: *Sign the paper.* Jassim refused each time. He was placed in solitary confinement in a nine-square-meter prison cell. The only source of light came from a small window covered with metal bars, high on the outer wall. Being alone in the cell was as mentally traumatizing as the torture was physically debilitating. He was aware of the fate of many of his colleagues, and prior to his capture Jassim had entertained thoughts of fleeing to Kurdistan, which was, if not entirely safe, out of Saddam's reach. He could then escape the country from there. Or maybe leave the country via Kuwait. In the end, however, he chose to stay in Baghdad and Chibayish, knowing that he might be arrested at any time. For Jassim, the choice was clear. If he went into exile, he might never see his family again. Or the Marshes either, for that matter. Staying was a risk he was willing to take. And it almost cost him his life.

Jassim knew many Communist Party members who took the easy way out and signed the letter disavowing the ICP and pledging allegiance to the Ba'th Party. Some were released, but others remained in jail and were tortured anyway. Jassim was also aware that in the eyes of the Ba'thists, the communist stain never went away. Former ICP members who were released continued to be suspects and were not accorded the same privileges as other Ba'th Party members. Jassim wanted nothing to do with the Ba'th Party.

After a few months of almost continuous abuse in Chibayish, Jassim was blindfolded, dragged out of his cell, and taken by van to the security center in Nasiriya, the capital of Dhi Qar Governorate. He was still wearing his blue dishdasha. Nasiriya was a larger city with a larger jail and an expanded arsenal of torture options. Prisoners were fed the same food, day after day, regardless of what jail they were in: watery lentil soup in the morning and then rice and bread in the evening. On occasion, they were given dates. In Nasiriya, Jassim was moved in and out of solitary confinement, sometimes sharing a slightly larger cell with five other Communist Party members who were subjected to the same treatment. His blue dishdasha was now in tatters. There was no contact with anyone outside the prison, and no way of knowing whether his family knew his whereabouts. Later, Jassim learned that six months after his arrest, a distant relative informed his family that he was still alive. Apart from that, they received no news.

The demand was the same for all the political prisoners. *Sign the paper. Cease your partisan activities. Join the Ba'th Party.* And the answer—for Jassim at least—was always the same: *No.*

The torture continued. The preferred method in Nasiriya was to blindfold the prisoner, hang him upside down from the ceiling, and batter him with a long rubber tube filled with small stones. Ribs were bruised; ribs were broken. At some point, Jassim would pass out and then wake up in the infirmary or on the floor of his cell. After a few days' rest, there would be another beating. Time had lost all meaning. Not knowing from one day to the next when the next round of torture would occur. Almost always alone. Always in pain. The expectation of another round of torture weighed heavily. By this point, he doubted he would survive. But his resolve never wavered: he was not going to join the Ba'th Party.

It was a shorter stay in Nasiriya. Some weeks. Six, seven, maybe eight. One day, he was blindfolded, taken from his cell, and shoved into the back of a van. A few hours later, the van stopped, and his blindfold was removed. Jassim was escorted into a large prison in the center of Baghdad, part of the Public Security Building. At this point, he had trouble walking. The only benefit of being transferred to Baghdad was that the jail was crowded with other political prisoners, and Jassim no longer had to endure solitary confinement. But this didn't stop the torture. His body was badly bruised from the beatings with stones, his legs and feet were covered with slashes and burns, and he had lost a considerable amount of weight. There didn't seem to be much else the security services could do to him—except there was.

A week after arriving in Baghdad, Jassim was again blindfolded and escorted by two guards to a small room. They pulled his arms in front of his body and then bound his wrists. Slowly, deliberatively, the guards rolled newspaper around his fingers, one at a time. Then they tied his fingers together with a thin piece of leather and sprayed them with oil. Hands held out in front of him, a moment of quiet. He remembers two sounds: the click of the lighter, and his screams.

Nine months after he was arrested in Chibayish, Jassim was released from the Baghdad jail, inexplicably. This was no cause for celebration; he was physically frail and mentally damaged. And still in his blue dishdasha. On his release, he was given a small amount of money—just enough to buy a new set of clothes and take public transportation, first to Nasiriya and then to Chibayish. He went into hospital for a few days and then was nursed back to some semblance of health by his family. They told him it was a cousin from Nasiriya, Jasseb, who helped get him released. Jasseb was a senior Ba'th Party member in Dhi Qar who disagreed with the torture and killing exercised by the security services, and had disdain for many of the decisions of the Ba'thists. He was eventually expelled from the party, but not before he facilitated Jassim's release.

Although still feeble, Jassim was encouraged by his family to return to Baghdad and continue his engineering studies. It wasn't without risks, and he continued to be harassed by the security services, who would bring him in for questioning and threaten him with beatings. But no arrest. Most of his friends and colleagues had left the country or been killed. Baghdad had changed; gone were the art shows and the poetry readings. According to the Ba'thists, they attracted a troublesome and undesirable element of society, and were therefore banned. The party of Saddam promoted modernism, but disdained any form of intellectual, cultural, or political activities, unless they benefited the party. Jassim also became disaffected with the Communist Party. The leadership had dissolved, and the party did little to help him while in prison. He relinquished his party membership in 1979. In 1980, Jassim graduated from university with a degree in machine engineering, but the timing was inauspicious because he was immediately drafted into the army. The Iran–Iraq War had just begun.

Al-Nasir Salah al-Din Yusuf ibn Ayyub, or Saladin for short, is best known in the West for leading military campaigns against the crusader states in the Middle East in the late 1100s. Saladin, a Sunni Muslim of Kurdish descent, was the first sultan of the Ayyubid Dynasty, which he founded in 1171. By this time, Saladin had already conquered Egypt and attacked crusader districts between there and Damascus. At the peak of the dynasty, which lasted for almost a century, Saladin ruled over seven million people and controlled Egypt, Syria, Turkey, Yemen, and parts of Saudi Arabia and North Africa. His greatest military accomplishments came in 1187, when he won a decisive victory over crusader armies at Hattin near Lake Tiberias (Sea of Galilee) and succeeded in capturing Acre, Nablus, Jaffa, and other cities in the region.

He then set his sights on Jerusalem, which by then was overrun by refugees from elsewhere in the region and had a reduced number of crusader knights, most having already surrendered to Saladin. Still, numerous

Figure 13. A traditional *mudhif* (courtesy of Jassim Al-Asadi).

attempts to take Jerusalem by force proved unsuccessful, and Saladin laid siege to the city. After six days, Balian of Ibelin, the crusader in charge of Jerusalem's defenses, agreed to a peaceful surrender. In the aftermath, both Christian and Muslim shrines were preserved, which was a testimony to Saladin's understanding of the historical importance of the city.

A key factor in these two major successes was water. Saladin ensured his armies had ample fresh water while simultaneously preventing his opponents from accessing additional water from outside the city. More than 700 years later, this strategy was repeated by Arab armies when they cut off water supply to Jerusalem in 1947 during the Civil War in Mandatory Palestine, thus demonstrating that water is not only essential to life on Earth but has strategic value as well.

Saladin wasn't the first to exploit the strategic value of water in the region, as demonstrated by the Entemena clay cone, which dates from 2400 BCE and has text engraved around its perimeter. It was found on the banks of the Euphrates near the ancient city of Babylon in the 1800s and now sits in the Louvre Museum in Paris. Inscribed on the cone is the oldest recorded story of a border conflict, between Umma and Lagash, two Sumerian city-states located on the Tigris River in southern Mesopotamia. At that time, the main channel of the Tigris was the present-day Gharraf River, and Umma was located roughly thirty kilometers north and upstream of Lagash. The two city-states engaged in numerous conflicts involving land and water between 2500 and 2400 BCE. The text on the clay cone credits Enlil, the Sumerian god of wind, air, earth, and water, for trying to ameliorate these conflicts by clearly demarcating the border between Umma and Lagash. Mesalim, the king of Kis, a nearby city-state, then set in place monuments (stele) to mark the border. However, there were transgressions on both sides.

Tensions reached a peak, and the leaders of the two city-states, King Ush of Umma and Eannatum of Lagash, engaged in battle. Eannatum, who was inspired into battle by the god Ningirsu—the patron god of Lagash and the god of agriculture, healing, hunting, law, scribes, and war—defeated King Ush. With Ningirsu by his side, there was little doubt of the outcome. As part of the spoils of war, Ningirsu and his earthly successors extracted heavy fines from Umma. Over time, the debt load became too great, and Ush, the new king of Umma, decided to discontinue payments.

Not long after, Entemena became king of Lagash and refused to forgive the debt and demanded payment. King Ush responded with a strategy that would guarantee his legacy in the annals of history. He diverted the Tigris River away from Lagash, stopping the flow of water downstream. This caused massive crop failure, resulting in economic and social hardship for Lagash. In an additional fit of rage, King Ush smashed all the border monuments. Ultimately, however, Entemena responded by attacking and defeating Umma and killing King Ush. The story inscribed on the clay cone ends by praising Entemena and promising curses against anyone who violates the original boundary.

Regardless of its veracity, the story indicates that water can be a strategic resource—more so in regions where it is scarce. An asymmetry in military power might reduce the vulnerability of those located downstream, but there is little doubt that being unable to access water, whether in Lagash four thousand years ago or anywhere in the world today, has profound health and economic implications. Not long after the battles between Umma and Lagash, the entire Akkadian Empire, of which Lagash was a part, collapsed abruptly. The likely cause was a major drought that affected all of Mesopotamia. Again, water was the issue. Water dictates where and how people live; it engenders intense feelings and can be a factor contributing to conflict. Water also does not recognize political boundaries.

The Tigris and Euphrates Rivers that emerge like two ribbons of water from the mountains of eastern Turkey and flow south through Syria and Iraq are vitally important to all three countries. In a water-scarce region, the two rivers offer a major source of drinking water for people and animals. They are also used to produce electricity, to provide much needed water for irrigating crops, and to act as a repository for waste. The Tigris and Euphrates Rivers are the lifeblood of the Iraqi Marshes. Iran, too, is a part of this discussion, since important tributaries that replenish the Tigris originate in Iran. All this then raises the question: Who owns the water?

In 2010, the United Nations declared that access to safe, clean drinking water and sanitation is a human right and is essential to the full enjoyment of life. Accordingly, water is owned by the Earth, and every human being has a right to safe and clean water. Assuming water is a community resource, it should be relatively easy to develop laws pertaining to its use—at least this might have been the case if water stayed in one place. But water moves, sometimes over long distances. In the process, it crosses political boundaries or may form the border between political jurisdictions. The Euphrates River flows for 2,800 kilometers, the Tigris for 1,850 kilometers. If two city-states thirty kilometers apart can fight a war over water, one can only imagine the issues that arise when water crosses international boundaries.

Not long after the En-metana cone was inscribed, laws appeared regarding conflicts over the use of water. Initially, these focused on ensuring that no one caused floods that might affect another's property. The Sumerian Code of Ur-Nammu (2100 BCE) included one water law: "If a man floods another man's field, he shall measure out three kur of barley per iku of field."[5] The Code Law of Ur-Nammu addresses the accidental or deliberate flooding of an adjacent field, an emphasis reinforced in the Code of Hammurabi some 350 years later.

Hammurabi was the sixth king of the First Babylonian Dynasty and reigned from 1792 to 1750 BCE. Although he expanded Babylonian rule to most of Mesopotamia, Hammurabi is perhaps best known for issuing what is, at present, the oldest deciphered set of laws. Hammurabi's Code consists of 281 laws, many dealing with physical punishments accorded to those found guilty. When people think of Hammurabi, they mostly recall the adage, "An eye for an eye, a tooth for a tooth." The actual deciphered wording of the law reads: "If a man put out the eye of another man, his eye shall be put out."[6] Hammurabi claimed to have received the laws from Shamash (also known as Utu), the sun god and the god of justice, morality, and truth. Four of Hammurabi's laws mention water specifically, and all four cases deal with the responsibility to ensure that floods caused by digging canals and irrigation works do not accidentally or purposely affect another person's property.

It was not until the fourteenth century BCE that laws addressed the type of action that Umma promulgated against Lagash; that is, whether an individual or a city-state could reduce the flow of water to the detriment of downstream users. The Hittites of Anatolia (now Turkey) developed a set of laws that recognized the responsibility of upstream users of water to protect downstream users from floods. They also prohibited robbing a downstream neighbor of water through damming or diversions. These laws were the first to recognize that water is not just a tool for economic development but can be used for oppugnant reasons as well.

Somewhat surprisingly, there is no mention of water laws in the Christian Old Testament or the Hebrew Bible. There are, however, a few stories that imply an inclination toward customary law, which is the process of following accepted rules of conduct rather than having a formal codified law. The story of Jacob and Rachel in Genesis (29: 1–11) illustrates that water was considered a common property resource. When Jacob made his trip from Beersheba to Haran, he met shepherds who were waiting to water their sheep from the community well. Access to the well water was controlled by placing a stone over the well, too heavy for one person to move. Once all the sheep were gathered, the shepherds would collectively move the stone to access the water. When Rachel arrived with her father's sheep and needed water, Jacob drew on unusual strength to remove the stone and give water to Rachel (whom he eventually marries). Implied in this story are two customary laws: one that pertains to community access to the resource, and the other that allows for sharing with those in need. But there is no specific law mentioned in the Bible about the use—or misuse—of water.

Unlike the Old Testament and the Hebrew Bible, the Qur'an has numerous references to water and how it should be used. In general, Islamic law considers water to be beyond private ownership; it originates with God and belongs to His community. No person can appropriate a river, for example, and no one can sell water.[7] Water must be shared, although Maliki, one school of Muslim law, gives individuals broad rights of ownership. In economic parlance, water is considered common property, whereas in cases where water is scarce, the Prophet decreed that priority should be given to those who live closest to the water. It also depends on whether the water is flowing or in place (such as a spring or a well). Water that flows in a river or stream can be used by adjacent landowners but then must flow to others downstream, much as expressed in the Hittite laws. Water that does not flow, on the other hand, is under the jurisdiction of the owner of the property, although owners are still required to share water with those in need. Conservation of water is also addressed in the Qur'an, in that people are expected to be frugal in their use of water and not pollute it.

The Qur'an does not address the issue of whether water must be shared with one's enemies. However, according to hadith—traditional law based on the words and actions of Prophet Muhammad—water can be used as an element in warfare. Once Jerusalem was conquered, Saladin provided water for the captured refugees, slaves, and citizens of the city, in accordance with Islamic law. Indeed, Saladin was following hadith in this respect. His actions may have also been influenced by the story of King Ush long before him, but they were also following hadith.

Figure 14. Women returning to their village with boats laden with reeds (courtesy of Jassim Al-Asadi).

The lack of a codified law regarding the shared use of water resources between countries or even within countries makes it easier for upstream riparians to claim they have the legal right to use water as they see fit, even if it negatively impacts those downstream. Despite the admonition by the UN that water is a human right, every decision regarding water becomes strategic. With their vulnerable location downstream of almost everything in Mesopotamia, the Iraqi Marshes became an unsuspecting and unwilling victim in a multitude of political, economic, environmental, and cultural conflicts over water that began in earnest in the latter half of the twentieth century and continue today.

3

WAR IN THE MARSHES

They proceeded slowly and apprehensively as Mihyar tried to grasp the scene he had fallen into like a child falling from a plane above the sea, not knowing where he was or how he would get out. He was amazed by the boundless expanse stretching forever, as the boat advanced through this morning whose sun was hidden by rushes rising up like *afrit* and *djinn* in the tales of the *One Thousand and One Nights*.

Haidar Haidar[1]

For many, living in Iraq under Saddam's regime was like being in prison, with few being allowed to enter or leave the country. Doctors, engineers, professors, and university graduates, among others, could not easily cross the border, and people with family or friends abroad were harassed. All political parties were banned, except for the Ba'th Party. If one publicly disagreed with government policy, both the individual and their family were at risk.

During Jassim's first two years at the University of Technology, his girlfriend was a classmate named Wesal. She lived with her parents and four siblings near the Al-Mustansiriya University campus in Baghdad. The youngest child was five years old and mentally disabled. The entire family had long been active in the Iraqi Communist Party, and Jassim would often spend evenings and nights in their home. Their values and interests aligned with his, and evening discussions on national and global issues with the family provided a welcome break from his technical studies. Jassim's relationship with Wesal and her family became stronger with each passing week. It all ended abruptly, however, when he was arrested in Chibayish in late August 1978.

When he returned to university after a year of incarceration and recovery, Jassim immediately went to see Wesal and her family. Knowing what happened to many communist party members during this period, Jassim worried that security forces might be occupying—or at least watching—the house. He approached cautiously, walking on the opposite side of the street. The house appeared empty, and so he carried on his way. Two days later, Jassim returned. This time, he saw clothes hanging from the window. He hesitantly walked up to the door and knocked.

"Who is there?" was the tentative response from a woman on the other side of the door.

"Just a friend," replied Jassim.

"Yes, you may be a friend, but who are you?" asked the woman.

Jassim hesitated. Finally, he answered. "It is Jassim."

The door opened slowly. Peering from inside was Wesal's mother.

"Jassim, come in," she said. "Quickly."

The only other occupant was her disabled son, now six years old. After Jassim told her about his arrest, he asked where Wesal was. Her mother had no idea. She told Jassim that not long after he disappeared, the security police came to the house looking for her husband. Knowing they might be sent to jail for their communist activities, Wesal, along with her father, older brother, and sister, went into hiding. The security forces were angry and arrested Wesal's mother and her handicapped child. The two of them spent the next seven months in jail. After she was released, nobody was able tell her what had happened to the rest of the family. She was unsure whether they were captured or managed to escape and was afraid to ask. And although Wesal's mother remained hopeful, Jassim knew that it was unlikely she would see her family again. Not long after, she moved with her son to Babylon to live with her parents and Jassim lost touch with her.

In the spring of 2004, twenty-five years after seeing Wesal's mother, Jassim was having coffee at a café in Babylon and reading an Iraqi newspaper. One of the lead articles was about the discovery of a mass grave in the al-Karkh Cemetery in western Baghdad. The grave was found by investigators for the new Coalition Provisional Authority (CPA) that was governing Iraq. They also discovered a document in the administrative office of the cemetery listing people who were buried there, and accompanying the newspaper article was a partial list of names from the gravesite. Wesal and her sister were near the top of the list. Jassim now knew for certain what he had long assumed: the two women had been arrested—and likely tortured—killed, and buried in the mass grave. There was no listing for Wesal's father or brother. Although he had always feared the worst, Jassim still felt despondent. Too many people had died brutally and unnecessarily at the hands of the previous regime. It was all too reminiscent of reading about Tahseen's execution in the paper twenty-four years before.

Figure 15. Wesal, 1978 (courtesy of Fatima Jabir Shaabith).

When he arrived home, it was clear to his wife Suad that something had happened. Jassim sat down and told her the entire story. It crushed him.

Khalid Rasul al-'Am, director of the al-Karkh Cemetery in Baghdad, later acknowledged that he had registered the names of 993 execution victims that were buried in a mass grave in al-Karkh between 1987 and 2003. Most were political prisoners. This was one of 259 mass graves that were found by the Combined Forensic Team of the CPA in 2003–2004.[2]

There is no simple or logical way to describe borders between countries. There is an array of country shapes, with lines that squiggle and squirm, jig here and jog there, or go in circles. Admittedly, there are a few borders that appear as simple, straight lines; the British seem to have had a hand in most of these, although the decision on where to draw the straight line may not have been so simple. And there are a few borders that are not illogical. Rivers, for example, form a natural border that is often used to demarcate states. There are even a few borders that follow the geographic coordinate system of longitude and latitude. In fact, the border between the U.S. and Canada falls along the forty-ninth parallel . . . until it doesn't. But regardless of how they came to be, borders do matter. Umma and Lagash may have been the first documented border dispute, but wars have been fought over borders for all recorded history and it seems unlikely this will change any time soon.

The border between Iraq and Iran falls into the category of the not-so-simple and very illogical, except for the sections that seem to make no sense at all. In the Marshes, the border hardly mattered—at least that was the case until oil was discovered. The border between the two countries runs for approximately 1,500 kilometers, from a point in the north where Iraq, Iran, and Turkey meet, to the mouth of the Shatt al-Arab River in the south, where it empties into the Gulf (figure 2). There are three distinct sections of Iraq's border with Iran, and all three have been contested over the past 500 years. For the first half of its distance, the border appears like a piece of string that was wound much too tightly and started bunching up in little knots. From north to south, it diverts east, then west, then east, then a bit back to the north, and so on, until it approaches Kut. It then disentangles and becomes smooth, if not straight, for roughly 200 kilometers, with a bit more doodling toward the end of the section. Sixty kilometers east of Amara, the second section of the border is a series of four straight lines, through the Marshes and south to the Nahr al-Khayin River. It is as if someone decided to experiment with a straightedge. The last section of the border follows the Nahr al-Khayin until it empties into the Shatt al-Arab River. From there, the border between Iraq and Iran becomes the thalweg—the line of lowest elevation between two banks of a river—until it meets the Gulf.

Drawing the boundary between two states using a meandering river, or even a relatively straight one, seems to make sense, and there are many examples of this around the world. However, the paths of rivers can change, particularly in regions such as southern Mesopotamia. The main channels of both the Tigris and the Euphrates have shifted in the past. The city of Ur used to be on the banks of the Euphrates River, whereas now its former location is sixteen kilometers to the west of the river. Putting a border on land—in this case on the bank of the river—versus making it the thalweg, is more than simply an academic argument; it can mean the gain or loss of land.

The present border between Iraq and Iran resulted from different sets of wars and negotiations over the past half-millennium. It began with hostilities between the Ottomans and the Safavid Dynasty in Persia in the early 1500s. A century of on-again, off-again war resulted in the two parties signing the Treaty of Zuhab in 1639. This treaty demarcated the northern boundary between Ottoman-ruled Mesopotamia and Persia as somewhere between the Zagros Mountains to the east and the Tigris River to the west, although the exact location was not agreed upon. In the south, the border was disputed. The Persians thought that the Shatt al-Arab River was a natural border. The Ottomans, on the other hand, believed that Arab tribes on both sides of the river constituted a single unit, and therefore the border was east of the river. It made a difference since the Shatt al-Arab was the main shipping and transportation route from Baghdad to Basra and into the Gulf. The Persians already had the navigable Bahmanshir River, which branched off from the Karun and ran parallel to the Shatt al-Arab. The issue remained unresolved, but nobody yet cared about the middle section, between the Hawizeh Marsh and the Shatt al-Arab. It was an area of wetlands, desert, and very few people.

The area to the east of the present border from Hawizeh Marsh to the Gulf is part of the Iranian province of Khuzestan. At various times in history the northwest section of Khuzestan, adjacent to the middle section of the border mentioned above, was part of Mesopotamia and under Baghdad's rule. During these periods, it was populated by Arab tribes, including the Bani Asad. The province was conquered by the Safavid Dynasty in 1510 CE and integrated into the rest of Persia. Nevertheless, there were always a few Arabs who felt this land rightfully belonged to Iraq, although at the time it was less important than other sections of the border. The Treaty of Zuhab did not end the wars, but at least there had been reasonable discussions over where the border should be located, and it set the stage for future border deliberations.

It was not until 1843 that a boundary commission was created with the mandate to officially map the border. The commission consisted of four members, one each from Turkey, Persia, Britain, and Russia, with the last two acting as mediators. After extensive negotiations, the Second Treaty of Erzurum was signed in 1847. This new treaty drew on the basic framework laid out in the earlier Treaty of Zuhab, but this time carefully delineated the actual border, addressing specific issues of concern with respect to Kurdistan in the north and the Shatt al-Arab River in the south. A casual observer might have concluded that the negotiators had drunk far too much arak when they drew the border, but the result was the most comprehensive agreement ever signed between the two empires.

The key contentious issue for the Persians remained the southern border delineated by the Shatt al-Arab River. Much to their dismay, the treaty defined the border as the strip of land on the eastern bank of the river (the Persian side) rather than the middle of the river, or the thalweg. The Ottomans had been unyielding on the issue and were supported by both the British and the Russians. Persia still had rights of navigation on the river, but not ownership. It was an issue that would take another century and then some to be resolved. The middle section of the border from east of Baghdad south to the Shatt al-Arab (now part of northwestern Khuzestan) remained ill-defined. It was of little concern to any of the main parties to the negotiation, at least not until oil was discovered.

Including British and Russian mediators on the border commission was no accident. The two countries began meddling in both the Ottoman Empire and Persia in the early 1600s. For the British, the initial points of concern were accessing trading routes to India and protecting river traffic along the Shatt al-Arab River. The Russians had two goals: the first was to take control of northern Iran to ensure their own security, if not to expand Russian territory; the second was to eventually gain access to a warm-water port in the south. While wary of each other's motives, Britain and Russia worked together to undermine the authority of both the Ottomans and Iran for three centuries.

At the dawn of the twentieth century, Russia and Britain began playing a more active role regarding borders. The Constitutional Revolution in Iran between 1905 and 1911 provided Russia the excuse it needed to establish a military presence in the north of Persia. Under the auspices of protecting the public interest, they occupied the Caucasus, Georgia, Turkmenistan, and Tajikistan. Russia's ambitions in the region, along with a heightened concern over the security of British-owned oilfields in southern Iran, near the border with Iraq, were reasons enough for the British to position military troops in the south, near Basra. They were away from the reach of the Ottomans, but close enough to keep an eye on the oil. It was a wise decision, since the nearby Abadan oil refinery in Iran would be crucial to British naval vessels during and after World War I.

The treaties, agreements, and attempts at reconciliation over four centuries might have helped, but tensions between Persia and the Ottomans lingered, particularly with respect to border issues. By 1900, both empires were in decline. The Russians and the British used this to their advantage. In 1907, to prevent other countries from staking a claim to the region, they agreed to cooperate by dividing the combined lands of the Ottoman Empire and Persia into three sections for the purposes of controlling trade and influence, with Russia active in the north, Britain in the south, and a neutral zone in the middle. Following these discussions, yet another commission was created to resolve the border issue. The result was the 1913 Istanbul Protocol. Persia again wanted the issue of ownership over the Shatt al-Arab addressed. Britain was adamant that the river lay entirely within the boundaries of the Ottoman Empire. Nevertheless, British surveys showed that the Shatt al-Arab had shifted its course in the previous 400 years, which complicated the negotiations. A bit of horse-trading between the colonial powers resulted in Russia's acceding to Britain's wishes. The border in the south remained as specified in the Second Treaty of Erzerum, with the Shatt al-Arab entirely within Turkey's hands.

The northern and southern borders had been set, despite Iran's objections. The task now was to demarcate the border in northwestern Khuzestan, through Hawizeh Marsh. The British pushed to delineate this middle section with four straight lines. Iran did not want the Ottomans to have access to three rivers that flowed into Hawizeh Marsh (the main one being the Karkheh River), which were crucial for irrigation. A second issue was oil. The Anglo–Persian Oil Company was founded in 1909, and a refinery was built in Abadan, Iran, near the Shatt al-Arab. Although the first wells were in Masjed Soleyman, well east of the Hawizeh Marsh, the British suspected northern Khuzestan might have oil as well.

The first of the straight border lines started east of Amara and went roughly forty kilometers to the southwest, through the Marshes. This ensured rivers such as the Karkheh were entirely within Persia. The second line was almost the same distance but went directly south—again, through the Marshes. The purpose was two-fold: first, to ensure that the border did not meet the Tigris above its confluence with the Euphrates, and second, to keep possible oil fields in Iran, where Britain would have easy access to the refinery. Had the border line continued south, it would have cut off the city of Basra, which straddled the Shatt al-Arab. As a result, a third line was drawn perpendicular to the second one, going east for thirty kilometers. The border then cut south for another forty kilometers before hitting the Nahr al-Khayin River and the Shatt al-Arab. It is strange to see these four straight lines in the middle of a twisty, meandering border, but there was some logic to its design. It was much less contentious than sections to the north and the south—at least for the moment. A few Iraqi leaders still felt the country had claims on the

Arab sections of Khuzestan. Equally as important, the border cut across the Hawizeh Marsh. In time, this would have a significant impact on the entire eastern section of the Iraqi Marshes.

Following World War I, the issue over ownership of the Shatt al-Arab River resurfaced. Iran refused to accept the Istanbul Protocol, while the new country of Iraq followed its Ottoman predecessor and claimed jurisdiction over the river. Britain only cared about having unimpeded access to the river so it could reach its refineries in Abadan. Conflict over the river escalated, but all parties understood the importance of continued navigation on the river. In 1937, a border treaty between Iraq and Iran was signed, reinforcing the Istanbul Protocol. The border would remain as the eastern bank of the Shatt al-Arab, with Iran being granted full rights of use, along with greater access for anchorage near Abadan. The treaty, however, was never ratified. World War II and the changing political landscape intervened, and Iraqi–Iranian relations deteriorated. The border issue continued to fester.

There were improved relations in the mid-1950s between Iraq's King Faisal II and Shah Reza Pahlavi of Iran, including an agreement to settle their differences over the border. But these were derailed by the Iraqi Revolution of 1958. The Shah continued to claim that the boundary between the two countries in the south should be the thalweg of the Shatt al-Arab, which was countered with a claim from Iraq demanding that parts of Khuzestan be ceded to Iraq. In the end, pressure to acquiesce to Iran's claims on the river came from an unlikely source. Iran had been working with the U.S. to destabilize the Iraqi government by supplying arms and support to Kurdish fighters in the north, who were engaged in a civil war to gain independence from Iraq—a conflict that weakened Iraq. To counteract revolutionary fervor in Kurdistan, Iraqi President al-Bakr met with the Shah in Algiers in 1975 and conceded that the river boundaries on the Shatt al-Arab would be along the thalweg. It was a considerable concession and a stain on the career of a young Saddam Hussein, who was part of the negotiating team.

The 1975 Algiers Agreement, often referred to as the Borders and Good Neighborly Relations Treaty, had other provisions, including joint security measures on the border and a peace and mutual trust clause. In return, Iraq received what it needed most at that moment: the Shah abandoned all support for the Kurds in northern Iraq. After almost 500 years, it seemed that the border issue involving the Shatt al-Arab was finally over. But five years later, controversy over the border reemerged.

After graduating from university, Jassim was immediately drafted into the air force to work as an engineer at the Nasiriya Air Force Base. Had he agreed to join the Ba'th Party, he would have been an officer. The base was a noisy place, with Russian-made MIG-23 fighter jets taking off on bombing runs to Iran almost every day, and U.S.-made Iranian F-4 fighters returning the favor and dropping bombs on the Nasiriya base. Most of the time, the damage was minimal. Runways were destroyed and then repaired the next day. Jassim, along with other members of the Iraqi engineering corps assigned to Nasiriya, built barracks and hangers, constructed roads, and worked on the runways. While there was always danger from a misguided bomb or missile, the Iranian objective was to destroy aircraft and runways, not use their expensive ordinance on a young group of engineers.

The only time Jassim saw the front lines of the war while he was in the army was when he was sent to Kurdistan. It was a military tradition to send those who served in the military in urban centers far from the Iran–Iraq border to the battle lines for one month. Jassim was joined by a few other engineers from Nasiriya, and on arriving in Kurdistan, they were greeted by the sound of Iranian artillery bombardment on nearby Mount Hadid. The front lines were close. The problem was that nobody in the Iraqi army in Kurdistan had any idea what to do with a group of engineers, at least not for such a short period of time. They needed either fighter jets or battle troops. The engineers spent the thirty days fishing and swimming in a nearby lake and were eventually sent back home without seeing any action.

Three years into his air force service, Jassim married his cousin Suad at a large ceremony in Chibayish. Suad was six years younger than Jassim. Their wedding ceremony took place over three days, on the island where Jassim's family lived. When people arrived for the celebration, the men went to his Uncle Hashim's mudhif, a large reed structure that served as a meeting place and coffee house. In addition to tea and coffee, the men were served meat, rice, chicken cooked with spices and beans, and traditional *tannour* bread cooked in a clay oven.

The women attending the wedding, all wearing colorful clothes, were received by Jassim's father and other uncles at the family home. On the evening of the first day, Suad and her family hosted a henna party for women and children from nearby islands, friends from high school, and some of Suad's former teachers. They dyed the tips of their hands with henna, danced, and sang songs. Jassim's female relatives sprinkled candy on the visitors and passed out sweets to the children. The black singer Fadhil played the *rebab*—a two-stringed instrument that is related to the oud (lute)—and the *al-duff*, a skinned instrument attached to a large piece of wood, which looks much like a very large tambourine. He sang in a loud, joyous voice, surrounded by a circle of young girls who were also singing and dancing.

On the second day, Suad wore a white wedding suit with long, wide sleeves that Jassim purchased for her in Baghdad. Jassim was dressed in a black suit, white shirt, and bright red tie. The official ceremony was presided over by a senior member of the community who was a *sayyid*, meaning he was a direct descendant of the Prophet Muhammad. He accompanied Suad from her family's house to Jassim's. More than 200 people attended the wedding. As befits weddings anywhere in the world, it was an occasion with much gaiety and laughter. Children were everywhere and had a contest jumping from boat to boat, until one of them fell in the water. A year later, Suad and Jassim had their first child, a daughter they named Noor—a word that means light or illumination.

Jassim was discharged from the army in April of 1985, along with others who had survived at least three years of military service. He was glad to have his military service behind him and, for the first time in many years, felt free. He decided to stay in Nasiriya, grow his family, and work for the provincial government as an irrigation engineer on the Main Outfall Drain (M.O.D.).

The M.O.D. is a 565-kilometer-long shallow canal that drains irrigation water runoff from agricultural land and moves it south to the Gulf. It is one of the largest drainage projects in the world. The runoff is polluted by salts, fertilizer, and pesticides, and the M.O.D. transports this polluted water away from the area, preventing the accumulation of salt in the soils that negatively affects agricultural output. The M.O.D. is important not just to the survival of agriculture but to all flora and fauna in the region. The water moves mainly through gravity flow, supplemented by pumping stations, such as the main one at Nasiriya—the largest pumping station in the Middle East. At Nasiriya, the polluted water is pumped under the Euphrates River, through the Hammar Marsh, and eventually to the Shatt al-Basra Canal, which parallels the Shatt al-Arab and flows into the Gulf.

Jassim worked for nine years on the M.O.D. It was an important project, with enormous potential benefits for the Marshes. Unfortunately, two months after he left the air force for his position with the Dhi Qar district government, Jassim was once again involved in the war. Not only did he lose his newly gained freedom, but he almost lost his life.

To Saddam Hussein, the 1975 Algiers Agreement was anathema; he felt that Iraq had given up sovereign territory for very little in return. When he became president of Iraq, he vowed to do something about it.

The Middle East experienced cataclysmic political change in 1979. Anwar al-Sadat of Egypt and Menachem Begin of Israel signed the Egypt–Israel Peace Treaty on March 26. While this was viewed by much of the world as a major step toward a broader Middle East framework for peace, Arab states were highly critical. Many broke off diplomatic relations with Egypt.[3] As a result of the agreement, Egypt lost its status as a leader in the Arab world, and Saddam saw this as an opportunity for Iraq to take its place.

Six days after the Egypt–Israel Peace Treaty was signed, Iran voted to become an Islamic republic, ruled by Islamic law. Ayatollah Khomeini had returned from exile two months previously, and by December would be named supreme leader of the country. This was perceived as a major threat to the secular, pan-Arabist Ba'thist government in Baghdad. After being exiled to Turkey in the early 1960s, the ayatollah moved to Iraq, where he spent fourteen years in Najaf and Karbala, stirring up dissent, criticizing the Ba'thists, and promoting his dream of an Islamic state before Saddam deported him to France in 1978. Antagonisms between the two men played out on the world stage even before the ayatollah returned to Iraq. Saddam wasn't the only one concerned; other Arab states became alarmed at how Iran might influence Islamic groups in their own countries. Much of the rest of the world started viewing Iran as a pariah state and a potential threat to the world order.

Saddam Hussein became president of Iraq in July of 1979, although for all intents and purposes, he had been running the country for the previous five years. Despite his efforts to strengthen the military, threats within the country remained troublesome. The Kurdish issue in the north had been partially contained when the Kurds lost their main sponsor, Iran, because of the Algiers Agreement. Many Kurdish leaders left the country, and Saddam ordered a massive relocation of Kurds out of the north. Nevertheless, they still posed a threat.

The Iraqi Communist Party (ICP) continued to garner popular support, both in Baghdad and in the south. The desultory relationship between the ICP and the Ba'thists turned sour even before Saddam became president, and he accelerated this with a brutal campaign of repression against the leadership. While the communists had been neutralized, al-Da'wa and the United Iraqi Alliance, the two largest Islamic political parties, were gaining popular support. Protests by Shi'a groups in Najaf and Karbala in 1977 raised concern within the Ba'th Party and, as was the case with the ICP, arrests and executions followed. The establishment of an Islamic state on their doorstep emboldened these Iraqi Islamist groups, and soon after assuming office, Saddam decreed that being a member of al-Da'wa was punishable by death.

The two world superpowers at the time, the United States and the USSR, did not escape from the tumult in the Middle East in 1979. In early November, a group of Iranian students in Tehran protesting U.S. support for the Shah (who by then was in exile and receiving medical treatment in the States), broke into the U.S. Embassy compound and took fifty-two hostages. It proved to be a death knell for U.S. President Jimmy Carter, who was voted out of office a year later. The animosity toward Iran was now firmly embedded in the U.S. consciousness. Saddam could not have planned it any better.

In the waning days of 1979, while the U.S. was preoccupied with Iran, the USSR staged a surprise invasion of Afghanistan, installing a Soviet loyalist as head of the country. It was a costly move for the USSR, both economically and in terms of lives lost, and it was eight years before they were able to remove their troops in ignominious defeat. During this rather chaotic time, Saddam Hussein saw an opportunity. The Iranian army was weakened and in disarray with the transfer of leadership in the country. In contrast, Saddam had been rearming and strengthening the Iraqi military for the previous five years, drawing on his close ties with Russia to purchase weapons. He purged both the military and the Ba'th Party of anyone who might be a threat to his leadership, promoted himself as the new champion of Arab nationalism, and traveled to other nations and developed a network of support around the Arab world.

Despite the Good Neighborly Relations Treaty of 1975, border skirmishes between Iran and Iraq continued and relations began to deteriorate.

By 1980, Ayatollah Khomeini was calling on the Iraqi public to revolt against the Ba'th regime. When Iran granted asylum to the Barzani brothers, two Kurdish leaders most responsible for attacks in the north against the Iraqi government forces, Saddam reached his breaking point. Iran's economy was in tatters, its military weak; it was the perfect time to go to war. It would also confirm his position as the leader of the Arab world. In September of 1980, Saddam abrogated the Algiers Agreement, taking back control of the Shatt al-Arab River. The skirmishes intensified, and Iraqi air force engaged in pre-emptive attacks on Iranian military sites.

The Iran–Iraq War officially began on September 22, 1980, when Iraqi fighter jets attacked various sites in Iran, including Tehran. Unofficially, however, it had been going on for months. The attack was not a complete surprise to Iran, and they responded accordingly. Over the next four days, Iran targeted Iraqi military bases and flew low over Iraq's major cities, making certain that the local citizenry knew that they, too, were vulnerable. Both sides lost over two dozen of their best fighter planes, with the Iraqi Air Force sustaining the most damage. Saddam underestimated the speed and ferocity of the Iranian response and was forced to temper his expectation of a quick victory. Many Iraqi citizens were already asking: Why go to war with Iran? The answer might have been clear to Saddam, but for the public, it was much more ambiguous.

Ostensibly, there were four reasons for the war: first and foremost—if one were to believe the public pronouncements coming from the government—was a dispute over the border; second was the Kurdish issue in the north; third was the personal animosity between Saddam and Ayatollah Khomeini; and fourth was a clash of ideology between the two regimes, with Iran's Islamic fundamentalism being perceived as a threat to Saddam's Iraq. The border issue may have been the most contentious of the four. Three sections of the border were contested: the primary one was along the Shatt al-Arab River, involving the long-standing dispute over whether the border between the two countries should be the eastern bank of the river or the thalweg; the second area was in Arabic-speaking western Khuzestan, near the Hawizeh Marsh; and the third was a bit farther north, northeast of Baghdad, in the foothills of the mountains—a small Arabic-speaking region that was promised to Iraq as part of the Algiers Agreement in 1975. Saddam was clear: He had been forced to agree to a change in the border along the Shatt al-Arab as part of the Algiers Agreement in return for

Iran's commitment to stop supporting Kurdish separatists in the north—and now he wanted it back. Iran was expected to meet other conditions as part of the agreement. In Saddam's opinion, they had not done so, and he felt justified abrogating the agreement.

The second reason for the war was the ongoing fight for a separate Kurdish state in the north. Although Iran cut back their involvement after the Algiers Agreement, they had once again begun supporting Kurdish rebellion as a way of undermining the regime in Baghdad. Without Iranian support, the Kurds could not survive. From Saddam's perspective, this had to stop.

The third ostensible reason for war was the intense personal animosity between Saddam and Ayatollah Khomeini, with vituperate verbal attacks between the two intensifying in the months leading up to the war. Still upset from being deported from Iraq in 1978, the ayatollah spoke of the "infidel" Ba'th Party, and had been calling for the overthrow of Saddam ever since departing the country. "Iraqi people should liberate themselves from the claws of the enemy," he stated. "Wake up and topple this corrupt regime in your Islamic country before it is too late."[4] The relationship between the two leaders was to be a point of contention throughout the war, as the ayatollah refused to negotiate an end to the war until Saddam was killed or banished from Iraq. It provided the perfect setting to initiate a major conflict.

Coupled with the animosity between the two leaders was a fundamental clash of ideology between the two regimes. The Ba'th Party was non-sectarian and staunchly Arab nationalist, and the Shi'a in Iraq, who constituted sixty percent of the population, were wary about being part of a united Arab world that was predominantly Sunni. The Kurds in the north, with another 20 percent of Iraq's population, were also opposed. The Islamic Revolution in Iran, a country with close ties to the Shi'a population in Iraq, represented a significant existential threat to the regime; there was justifiable fear in the Iraqi government that Iran would export their revolution to Iraq. The ayatollah's continued rantings about the need to overthrow Saddam and the regime represented the public face of this threat.

The Iraqi government responded initially with brutal repression against al-Da'wa and other Islamist parties, as well as the Shi'a leadership. These actions were countered by protests in the Shi'a holy cities against the government. Saddam viewed the war as a battle against religious fundamentalism—in favor of a different kind of fundamentalism—and a way to stop Iran's involvement with both the Shiʻa and the Kurds. He knew that most other Arab states felt the same and that they would support him if he went to war.

An additional incentive for the war was Saddam's intense desire to be not just the leader of Iraq, but the grand leader of the entire Arab world. This would provide Iraq with a platform for regional hegemony in the Middle East. He viewed Iraq as the cornerstone of the region, and failure for Iraq in its war with Iran meant the failure of the entire Arab nation. The agreement between Israel and Egypt was, for Saddam, fortuitous; it provided him the opportunity he needed to make Iraq the center of the Arab world. Syria, under Hafez al-Assad, was a likely contender for the leadership role as well, and Saddam wanted to demonstrate that he not only was a better leader but also more committed to Arab nationalism. A short war that brought Iran to its knees would be the perfect solution.

Arab states were mixed in their reactions and support for the war. There were important historical, cultural, ethnic, and religious differences among countries that affected their relationship with both Iran and Iraq and influenced how they viewed the war. While most countries with large Sunni majorities did worry about the rise of Islamic fundamentalism, they were not ready to unreservedly embrace Saddam's secular policies, or his leadership. War was an expensive proposition, even for two countries with large amounts of oil. Iraq needed financial assistance to keep its army and air force supplied with weapons and to keep its economy running. This was more important after oil production facilities were damaged by the fighting. Accordingly, it courted Kuwait and Saudi Arabia. Initially reluctant, both eventually provided substantial funding for the war. By 1988, when the war ended, these two countries alone had loaned Iraq $40 billion. With oil prices low and production affected by the war, Iraq's inability to pay off this debt would have troublesome consequences.

Regardless of the plethora of possible causes of the Iran–Iraq War, one thing is clear: there was no popular support in Iraq for the war. Even so, the state-run media ran an effective propaganda campaign aimed at convincing the public and the outside world that Iraq was fighting a war on behalf of the entire Arab world to stop the spread of Islamic fundamentalism—a war over grievances and loss of land that had accumulated over the years. Rather, the war was a misguided and self-centered attempt by

Saddam to elevate himself to the position of leader of the Arab world, and Iraq to regional dominance. Most of all, it was a costly miscalculation, both in economic terms and in loss of life. It was also a war that greatly impacted the Marshes.

There are many reasons why armies avoided fighting in the Marshes. While mythical demons ('afarit) and supernatural creatures (jinn) may be among them, the difficulty in traversing the mud, reeds, and water, the impossibility of bringing horses into the wetlands, and the ability of the residents of the Marshes to foment guerilla warfare and then disappear, all made conquering the region virtually impossible. The tribes of the south were formidable adversaries—better to stay on the edge of the Marsh and let well enough alone.

When tribal chief Marduk-apla-iddina II (known as Mero-dach-Baladan II in the Christian Bible), leader of the Chaldeans, revolted against the Assyrian Empire in 694 BCE, King Sennacherib mounted a campaign against him and defeated the Chaldeans. Marduk, however, escaped into the Marshes. Sennacherib followed with his troops, and while they destroyed adjacent fields and took over 200,000 people as slaves, along with horses, mules, cattle, and sheep, they were unable to penetrate the Marshes and never found Marduk. Five years later, Sennacherib tried again, this time taking boats down the Tigris and into the Marshes in hopes of capturing Marduk. A great flood drove him back, and the Marshes once again proved victorious.

Other armies attempted to tame the local tribes. The Abbasids, who ruled Iraq from roughly 1250 BCE to 750 BCE, the Persians, who followed and laid claim to much of southern Iraq, and the Ottomans, who ruled Mesopotamia for over 400 years leading up to World War I, all tried to penetrate the Marshes. It was always in vain.

One issue that did have larger implications was tribal control of the Tigris River. The river was navigable from Baghdad to Basra, and this afforded an opportunity for the Bani Lam, one of the more powerful tribes in the region, to pillage goods or extract fees for passage along the river. The Ottoman pasha in Baghdad would then send troops to control the Bani Lam and kill the leaders, but most would melt back into the Marshes and reappear when things quieted down. After the Persians invaded Basra in the late 1700s, the head sheikh of the Muntafiq—a confederation of tribes on the lower Euphrates—lured the Persian army of 12,000 men into the Marshes. When the Persians got stuck trying to navigate the water, reeds, and mud, the Muntafiq slaughtered all but three of them.

The Bani Asad, who by this time had settled in Chibayish, also proved a formidable adversary for the Ottomans. From a military perspective, the Marshes were a region best to avoid. On the other hand, the Marshes could provide a strategic advantage in battle, as the tribes had demonstrated. Saddam was aware of this as he set his sights on war with Iran.

In the 1970s, twenty villages lay adjacent to Umm al-Ni'aj, the largest lake in Hawizeh Marsh, each with a population between 500 and 1,500. People living on the perimeter of the marsh grew wheat and barley, and harvest time was in spring and summer. Other residents engaged in fishing or raising buffalo, and most hunted waterfowl in the winter, since Hawizeh was a popular wintering spot for migratory birds. Some residents created handicrafts for sale in local or regional markets. It was a quiet, rural existence, far from large urban centers—a region soon to be devastated by war.

Khaled Assam came to love Hawizeh, despite his initial experience there. He grew up in al-Dawaya, eighty kilometers north of Nasiriya. 'Awina, a small marsh of less than twenty square kilometers was close by. Fed by small tributaries to the Gharraf River, the tiny marsh was isolated from the Central Marsh and vulnerable to drought. Khaled felt a strong affinity to the Marshes, which were a stark contrast to the hot, dry, and dusty towns and regions to the west. He spent as much time as possible in the 'Awina Marsh. When Khaled came of age in the mid-1980s, however, it was a rude awakening. He was drafted into the army and posted to the Hawizeh Marsh, along the border with Iran. At first sight, he found it an enchanting place. The expanse of water in the lake and the dark green vegetation contrasted with the vastness of the blue sky, with birds everywhere, as well as fish and lichens. The local people survived as they did thousands of years before, building with local materials and living in small villages at the edge of the marsh. Things, however, were about to change.

Marshes and wetlands were a major impediment to an advancing Iranian army, but they could pose transportation problems for Iraqi forces as well. Khaled watched as the lagoon and marshes became mottled with patches of dry land that could withstand the weight of tanks moving to the front. The lush green reeds on the horizon were replaced by soldiers and killing machines. The transformation to a dystopian landscape was unsettling for Khaled, and although they would not meet until many years later, Jassim was one of the engineers working in Hawizeh, contributing to that transformation.

The Iran–Iraq War spanned eight years, making it the longest conventional war of the twentieth century. By the time the war ended in 1988, the two countries sustained over one million casualties, and the total cost of the war—including direct and indirect damage—was estimated at over one trillion dollars.[5] Ground battles took place all along the border, from Kurdish areas in the north to al-Faw, the southernmost point in Iraq. Some of the most intense fighting occurred along the Shatt al-Arab River, near the cities of Basra and Khorramshahr. The Marshes, however, were not immune to the fighting and devastation. Hawizeh Marsh, which straddles the border between Iraq and western Khuzestan in Iran, was directly involved.

The main source of revenue for both Iran and Iraq in 1980 was oil. Without it, neither country could have afforded a war. It therefore made sense that one of Iraq's wartime strategies was to attack Iran's oil sector. Thus, Iraq focused its efforts on damaging Iran's downstream oil infrastructure, such as large tanker ships and refineries. These were mainly along the Shatt al-Arab River and south to the Gulf—very accessible for Iraqi fighter jets. The problem with this strategy was that Iran had much the same idea, except it was looking to target Iraq's oil production facilities. This made the Majnoon oilfield in the southern Hawizeh Marsh vulnerable.

During the first two weeks of the war, Iraqi forces occupied much of western Khuzestan, including Hor al-Azim, the portion of the Hawizeh Marsh that lies within Iran. The offensive soon stalled, however, due to an inadequate supply chain. Iran counterattacked and slowly pushed Iraq back toward the border. In late 1981, Iran set its sights on occupying Iraqi land, although it was not an ideal time for any initiative: the wet season had begun, marsh areas were reflooded, and conditions on dirt roads were unsuitable for troop movements, let alone tanks. On the other hand, the inclement weather also prevented the Iraqi Air Force from bombing Iranian positions.

The Iraqis were surprised by the attack, given the time of year and the weather, and were unprepared. It took a few days to bring in extra troops, and by that time the Iranians had advanced to the Iraqi border at the north end of Hawizeh. But they could go no farther, the rain and mud proving to be too much of a barrier. Although they did not manage to cross the border, Iran was able to regain 40 percent of the land in Khuzestan that they lost early in the war. From a military standpoint, it was the first time Iran sent waves of troops into battle, one after the other, to overwhelm Iraqi forces. The counter-offensive came at an enormous cost: over 7,500 Iranians were killed or wounded, with Iraq sustaining roughly half that many casualties. There was little doubt that the Iranians would attack again in the dry season, and Saddam decided it was time to enlist the Hawizeh Marsh in his war effort.

Khaled witnessed the transformation. Where possible, the Marshes were emptied of water, roads built and reinforced to allow for easy movement of heavy equipment, bunkers dug for advancing troops, and antitank and antipersonnel mines buried throughout the region. Reeds were cut down and the natural environment was desecrated. More work was done in southern Khuzestan, including building a wall around the captured city of Khorramshahr and reinforcing Basra. What started out as a short war with the promise of victory was turning out to be a long ordeal.

The Iranians launched their next offensive in July of 1982, during Ramadan. The dry season had come, and there was reason to believe Iranian forces would advance into Iraq, with the goal of reaching Qurna and taking control of the main highway from Baghdad to Basra. At the start of the battle, the Iraqi army gave way, enticing Iran's forces to enter the Marshes. The Iranian troops expected a dry marsh but were in for a rude awakening. Two weeks before, the Iraqi command opened irrigation channels and flooded much of Hawizeh, and now the Iranians were trapped in the quagmire and sustained heavy losses as a result. Indeed, after only two days of fighting, they were forced to retreat. As happened many times in the past, the Marshes had won.

The Iranians were not deterred, however, and continued to launch counteroffensives in the south, near Basra, and to the north in Kurdistan. Slowly, they pushed Iraqi troops back toward the border. By early 1984, Iran had removed Iraqi forces from all their territories, except for a small area near the town of Qasr-e Shirin, northeast of Baghdad. With Iraqi forces on the defensive and morale low, it was an opportune time for Iran to lay siege to part of Iraq and possibly bring an end to the war. The Iranian military set their sights on capturing the Majnoon oil fields in Hawizeh Marsh and then carrying on to the Baghdad–Basra highway. The Iraqis had different ideas. Their strategy was to keep Hawizeh full of water to deter future attacks. The additional water made the area virtually impenetrable for Iranian troops, and the Iraqi command saw little need to post many defensive forces in the region. This decision proved to be a mistake.

Aluminum boats, some that held up to 100 soldiers, provided the solution. Iranian troops moved quickly across the lake and the wetlands. In the north, they captured two unprotected small villages. Less than two days later, the Iranians reached the Baghdad–Basra Road, although they were poorly equipped to defend it; heavy armaments could not manage the trip across the Marshes. They had reached their goal but were unable to hold the road and had to retreat.

Just to the south lay the Majnoon Islands and the Majnoon oil fields. Surrounded by water and with production shut down, the facilities only had a few Iraqi troops defending them. Iranian forces captured the islands without firing a shot. The Iraqi command panicked and sent troops and tanks to the border, but they were victims of their own success in flooding the Marshes. The Iraqi army was forced to retreat. When the Iraqi tanks followed, they became stuck in the mud, and thirty had to be abandoned as a result. The Iranians remained in the Marshes, along the border, and in control of the Majnoon Islands.

The loss of the Majnoon oil fields was a major blow to Iraq, and the high command moved quickly to develop new strategies for how best to use the Marshes to their benefit. In little more than a week, they agreed on two actions: the first was to run a high-tension electric wire through part of the Hawizeh Marsh to electrocute Iranian troops moving through the water; the second was to spray Tabun nerve gas over the Iranians, who lacked gas masks and would be slowed in their retreat by the Marshes. A month later, the plan was put into action. Small planes sprayed nerve gas on the Iranian troops as they tried to hastily retreat through the Marshes. During the ensuing chaos, the Iraqi command flipped the electric switch. Over the next few hours, thousands died from being electrocuted, and thousands more suffocated from the gas. Others who tried to run away from the front were shot by their own security forces for desertion, and anyone who escaped the Marshes and reached dry land inside Iraq was confronted by Iraqi tanks. It was carnage.

The Iraqis had been successful in driving Iranian forces back to their side of the border, with one exception: they were unable to recapture the Majnoon Islands. The Iranians suffered 50,000 casualties in the First

Battle of the Marshes, while the Iraqis sustained 12,000. It might have been considered a victory for the latter, had Iraq not lost control of almost 20 percent of its oil reserves.

Iranian armed forces were being decimated in the war. Defending their territory early in the war and then mounting offensives against Iraq had taken their toll. Women were mobilized and trained to guard sensitive security sites. When he came to power, Ayatollah Khomeini established an organization of civilian volunteers aged eighteen to forty-five who would be part of the Islamic Revolutionary Guard Corps. The group was called the Basij (a word meaning mobilization in Persian). When the war began, thousands of volunteers joined the Basij, and their ages ranged from twelve to eighty. The Revolutionary Guard employed the Basij as part of their strategy to send waves of combatants toward the enemy. The first few waves would help clear minefields and draw fire from the Iraqis to help identify their positions. Almost half a million Basijis were sent to the front, and two-thirds of these were children under the age of eighteen who had been indoctrinated in school and influenced by the media to join and become martyrs for the cause.

Chemical weapons (eventually used by both sides) and electrocutions were pronounced deplorable by the United Nations and the international community. Nevertheless, the use of child soldiers by Iran might have been the most perverse and disturbing aspect of the war. Almost 60 percent of the Iranian troops in the First Battle of the Marshes were children between the ages of twelve and seventeen. Ten thousand were killed or wounded. They were Basiji who had been given two to three weeks of training and sent off to the front.

A year later, in early 1985, Iran renewed the offensive in the Second Battle of the Marshes. In a surprise attack, troops took the same path through the Marshes using aluminum boats. Their target was the same: the Baghdad–Basra highway, just west of the Tigris River. Forces also left from the Majnoon Islands and pushed west. The Iranians soon were in control of the highway and had captured the main bridge over the Euphrates at Qurna, which was key for Iraqi resupply. The Iraqis knew that they had to move the Iranians back across the border. The counteroffensive began with the extensive use of chemical weapons. This time, the Iranians came prepared with gas masks, which decreased the impact of the chemicals but also made it difficult to maneuver, particularly in the Marshes. The Iranians were soon overwhelmed and retreated over the border. In the process, Iraq took over 2,000 prisoners. Despite their success, however, the Iraqis failed to reclaim the Majnoon Islands. The Iranians were entrenched in their position and virtually unassailable.

Fighting intensified in the south, near Basra, and on the al-Faw peninsula, but the Marshes remained relatively quiet. Saddam worried that the 20,000 Shi'a residents around Hawizeh might revolt against him or, at the very least, would pass intelligence to the Iranians, even though they had no desire for war. His solution was to remove them from their villages and transport them by military trucks to a camp in al-Maimouna, a remote district north of the Central Marsh. Any male between the ages of eighteen and forty was forced into the army. Anyone who refused was shot. Former Hawizeh residents were not allowed to return to their homes until 1992, four years after the end of the war.

When Khaled was posted to Hawizeh, the destruction of the marshes was just beginning. The process of filling the Marshes and then drying them out wreaked havoc on the natural system, severely damaging vegetation and soil. He watched as beautiful wetlands were cut with a patchwork of dirt roads and then dried to salt flats so foot soldiers could easily advance in the event of an attack. Pumping crews flooded other areas, making them an obstacle for the enemy. Fragments of artillery and rockets were strewn about, a testimony to the bloody landscape. Because the area separating the two combatants was full of dense green reed forests, the Iraqi military cut it all down as part of a comprehensive campaign led by the first deputy of the Revolutionary Command Council. Vast areas in front of Khaled's unit were turned into dry, brown land, devoid of life. Khaled also observed the local population being forcibly displaced from their villages and agricultural areas and moved to designated spaces to the north where they remained under Iraqi security surveillance. All non-military activities ceased for 200 kilometers around the lake. There were no more buffalo or cows, and marshes were transformed to ruin before his eyes.

It was clear to the Iraqi command that recapturing the Majnoon oil fields would not be an easy task. The Marshes created an enormous moat around a castle of oil infrastructure; this made it almost impossible to cross in the face of enemy fire. After the Second War of the Marshes, the Iraqi command abandoned their strategy of flooding the Marshes and instead ordered parts of Hawizeh Marsh to be drained, and roads and bunkers built through the area. This would allow easy access for troops and heavy machinery. The devastation had begun.

Not long after arriving in his new job with the Dhi Qar regional government in Nasiriya, Jassim was forced back into the war for six months—this time working for the army. The Iraqi high command needed engineers to help with the construction activities on the front lines, particularly in Hawizeh Marsh. Much against his will, Jassim was transferred to the Majnoon sector, where he was given responsibility for the main workshop that repaired heavy machinery for the army—machinery that was used to construct earthen mounds and roads in the Marshes for military operations, including trucks, bulldozers, excavators, and dump trucks. Repair work took place during the day, but construction work in the Marshes occurred only at night, to avoid Iranian fighter jets.

Jassim's group was assigned to build roads through the Marshes to allow Iraqi military units easy access to combat zones around the Majnoon oil field. Working without lights, they transported soil from a site east of Majnoon to areas close to the Iranian army. The Iranians would fire flares in the sky to better see the operations; these were invariably followed by mortars and sporadic gunfire. It was an atmosphere fraught with danger, killing, panic, and fear. Whenever a new road was completed, another attack by Iraqi forces soon followed. And although many of Jassim's colleagues were killed or injured, the work continued. It was vital for the army to be able to move tanks and troops through the area without getting bogged down in the Marshes. The Majnoon oil fields were too important.

The lead official on the project was a civil engineer by the name of Sami Ajami who cared little about casualties, as long as the roads were built. When Jassim completed his requisite six months, Ajami ordered him to stay at the front, citing the need to complete the Majnoon project. Jassim refused.

"Do you want me to have you thrown in jail?" Sami asked.

"Jail is for thieves," Jassim countered.

"Do you mean that I am a thief?" asked Ajami.

"You know what is inside of you."

Jassim once again was fortunate to have another savior. Ajami's assistant engineer in charge, Mohammad al-Janabi, was a good friend of Jassim's.

Over an evening of whisky drinking, Muhammad convinced Ajami to give up the idea of sending Jassim to prison—better to be rid of Jassim's tongue and his stubbornness. The next day, Jassim was on his way home, first to Chibayish and then to Nasiriya.

The battle to reclaim the Majnoon Islands, which Saddam called Operation *Tawakkalna 'ala Allah* (we put our trust in Allah), was launched in June of 1988. The Iranians had 40,000 troops defending the area, and Iraq countered with a force of 160,000 men and over 1,000 cannons. By comparison, it was more troops than landed in Normandy on D-Day in World War II. Massive artillery shelling was followed once again by chemical weapons. It took less than a day for Iranian troops guarding the oil fields to flee back into Iran. Except for small pockets of resistance in Kurdistan, Iranian troops had been pushed out of Iraq. The war was almost over.

Iran launched a major initiative in early 1986 to capture al-Faw Peninsula, the southernmost point of land in Iraq. The peninsula was desolate and covered by saltwater marshes, but it was a strategic location. Controlling the peninsula meant controlling the mouth of the Shatt al-Arab River, along with much of the oil tanker traffic from Iraq and Kuwait to the Gulf, potentially affecting oil delivery around the world. After two oil supply disruptions in the 1970s, the West wanted to ensure it would not happen again. The Iraqis knew of the buildup to the south but always assumed the target was Basra. One million Iranians were enlisted for the battle of al-Faw, including 200,000 Basijis. Iraq once again countered with chemical weapons, but most of their force remained guarding Basra, and Iran soon controlled the peninsula. Saddam wanted to use Kuwait's Bubiyan Island, immediately across from al-Faw Peninsula, to launch counterattacks. Fearing this might affect their oil facilities, the Kuwaiti government refused and threatened to stop financial assistance to Iraq. Saddam capitulated but never forgot this slight. Skirmishes continued around al-Faw, but it would take two more years before Iraq would regain control of the peninsula.

Iran's victory at al-Faw galvanized much of the world. The threat of an Iranian victory in the war might destabilize the entire Middle East, and that was too great a risk. The United States, the USSR, and most European powers began supporting Iraq, providing them with needed arms and intelligence. The U.S. Navy began controlling the Gulf, assisting oil tankers from Kuwait and ensuring that Iranian ships and jets did not harass international shipping. By late 1987, over seventy American, British, and French ships patrolled the Gulf. Iran continued with its attacks on oil tankers. What had been a conflict between two countries—a conflict that the world was happy to see play out—now had the potential to escalate.

The UN Security Council unanimously adopted Resolution 598 in July of 1987, calling for an immediate ceasefire between the two countries, a return to the previous international border, and repatriation of all prisoners (by the end of the war, Iran held 70,000 Iraqi prisoners and Iraq had 40,000 Iranians). It also included the provision for UN peacekeeping forces to be stationed along the border. Refusal to abide by the resolution could result in sanctions on one or both countries. At the time, Iran seemed to be gaining the upper hand in the war, and they had little incentive to agree to the resolution. But intervention from the West began to make a difference: Iraq's superior weapons and improved intelligence capabilities translated into a significant advantage on the battlefield. Saddam was loathe to agree to a ceasefire just yet.

The human cost of the war was taking its toll on Iran. It was clear they were no longer fighting just Iraq, but much of the Western world. The escalation had at least one tragic, unintended consequence. On July 3, 1988, a U.S. Navy ship, the U.S. Vincennes, mistakenly entered Iranian territorial waters while engaging Iranian gunboats in the Gulf, thus exposing it to a legitimate attack from Iran. Its position put the ship directly under a commercial aviation flight path from Dubai to Tehran: Iran Air Flight 655 had just taken off from Dubai and was ascending to cruising altitude when it was spotted by the U.S. Vincennes. The ship's crew thought it was under attack and fired two missiles, destroying the plane and killing all 290 people on board. Eight years later, the U.S. paid over $130 million to Iran in compensation.

On July 20, 1988, the Iranian regime agreed to accept unconditionally the terms of UN Resolution 598, to take effect one month later. It began drawing its troops from Iraqi Kurdistan and brought an end to air operations. Iraq used the time before the ceasefire took effect to bomb Iranian oil installations and a nuclear power station that was under construction.

The war officially ended on August 20, 1988. Despite Saddam's claim that it was a great victory for Iraq, the human and financial costs for both sides were enormous. And it would be a long time until Hawizeh Marsh recovered. What had been a war in the Marshes soon developed into a war *on* the Marshes. While Saddam managed to stem the tide of growing Islamic fundamentalism in the south, he was unable to achieve a major goal of his war: to reestablish Iraqi control over the Shatt al-Arab. The border remained the thalweg, as designated in the Algiers Agreement.

Khaled Assam survived the war. He was injured a half-dozen times, but after every recovery, he always returned to the battlefield. He had no choice: threats and intimidation were powerful encouragements. Khaled lost many friends in the Marshes. Most were killed by the Iranians, but some were shot by Saddam, either because they abandoned their posts or because they were deemed threats to the regime. Khaled's most vivid memory, however, remains the destruction of the natural environment. Thousands of fish appeared like brown and tan writhing masses of slime in the shallow water before they died from a lack of oxygen. The birds fled, having lost their shelter. Frogs and turtles slowly moved to other areas, but few made it . . . An eight-year war had thwarted the dreams of so many and wreaked havoc on nature.

Twenty years later, Khaled would return to Hawizeh, this time as a border control agent. Once again, he would experience the devastation of the Marshes.

4

REVOLUTION, REPRISAL, AND
THE ASSAULT ON THE MARSHES

I rubbed the leaf of an orange in my hands
As I had been told to do
So that I could smell its scent
But before my hand could reach my nose
I had lost my home and become a refugee!
Mourid Barghouti[1]

T he most important part of Muhammad Diwach Al-Asadi's day was
meeting his friends in the evening for coffee at his cousin Hashim's
mudhif. The cousins lived with their families on an island in the
Central Marsh, roughly three kilometers from Chibayish. In the morning,
Muhammad paddled his mash-huf through dense patches of reeds and an
array of canals to his general store in town, for it wasn't until 1988 that a
road was built between the island and Chibayish. When he returned home
in the evening, Muhammad would have dinner and then join other members of his clan, along with an occasional guest, for coffee in Hashim's
mudhif. The mudhif served as a meeting place, coffee house, guest house,
and community center. While most mudhifs were owned by sheikhs (the
tribal or clan leaders), a few others outside of town, like Hashim, were
granted permission to build a mudhif because of their socially significant
lineage. In addition to managing the activities in the mudhif, Hashim
earned a meagre income by weaving mats from dried reeds which Muhammad then sold in his store.

The sound of coffee beans being ground with a metal rod was the sign
for Muhammad to enter the mudhif. Once inside, he would take a seat near

the front of the large reed structure, as befitted someone with high status in the Bani Asad tribe. The mudhif is available to both clan members and visitors. Hashim, as owner, would assign seating inside the mudhif and pre-side over the activities. His mudhif was much like those built a millennium before, the basic frame consisting of pillars of reeds that were secured in the soil on opposite sides and then pulled together at the top to form an arch. There were nine arches in the structure, the number of arches often indicative of the importance of the owner, and always an odd number, since it is believed that Allah preferred odd numbers.[2] The front and back of the mudhif were supported by four vertical pillars of reeds, and woven reed mats covered the roof, sides, and floor. As was the case with all mudhifs, the entrance faced toward Mecca.

Figure 16. A *mudhif* under construction (courtesy of Jassim Al-Asadi).

Figure 17. Inside a modern *mudhif* (courtesy of Jassim Al-Asadi).

On occasion, Muhammad would entice his young son Jassim to join him for evening coffee. The only light in the reed structure was from a small fire in the center of the mudhif that provided a modicum of heat in winter and kept the coffee warm, as well as a few candles strategically placed to illuminate individual faces. Patrons sat on embroidered cushions and were served coffee or tea. Cigarette smoke wafted to the top of the arched structure, over time darkening the reeds on the ceiling. These were social gatherings, and most nights there were ten to twelve people present.

By 1990, Jassim had become a regular at the nightly meetings in Hashim's mudhif. He had survived incarceration, torture, and the Iran–Iraq War, and was now back in Chibayish with Suad and his daughter Noor, as well as two sons: Samer, born in 1986, and Maher, born in 1988.

Jabbar—a teacher and one of Hashim's nephews—also frequented the gatherings. Jabbar was a valued guest because he invariably brought along his battery-powered radio. The radio would be positioned near the middle of the group, and the signal was strong enough to pick up a few stations, including the BBC, which was broadcast in Arabic from the United Kingdom. And although listening to the BBC was considered a suspicious activity by the Iraqi government, it was nevertheless a nightly tradition inside Hashim's mudhif to tune into the world news from the UK, since

Iraqi radio stations broadcast only what the government wanted people to hear. After the BBC broadcast, discussions would ensue about both national and world events. But not every night. One of those who regularly attended these evening gatherings had a son who worked for Saddam Hussein's security services. Whenever the son visited, the radio played Iraqi music, and the discussion stayed well away from politics.

The Iran–Iraq War left Iraq strong militarily, but weak economically. The country incurred a massive debt with other Arab nations for arms purchases during the war, and it badly needed funds for post-war reconstruction. Oil accounted for 95 percent of the country's export revenues in 1988. This, too, was a problem, since the price of oil, which had peaked at $35 per barrel in 1980, had declined to only $11 per barrel. Saddam pleaded with members of the Organization of Petroleum Exporting Countries (OPEC) to cut back oil production to raise prices and allow Iraq to recover; however, Saudi Arabia and Kuwait—the two countries that carried most of Iraq's debt—refused.

One of Saddam's first actions after the war was to restructure the senior military to ensure that he retained total control. Family members and his most trusted friends were put in positions of power, while anyone who might be considered a threat was either released or killed. He then implemented austerity measures across the government, closing departments and laying off staff, but this had a perverse economic effect and increased unemployment. Saddam began looking for other avenues to raise money, and Kuwait was an obvious one.

Relations between Iraq and Kuwait at the time were strained and had been since the latter became independent from Britain in 1961. Tensions only increased with the onset of the Iran–Iraq War. The reasons were many. Iraq claimed that Kuwait was engaged in slant drilling (drilling at an angle) to tap into its neighbor's oil field. Iraq cited evidence to show that output from their Rumaila oil field declined during the 1980s, while output from a well only a few kilometers away, inside Kuwait, increased. Kuwait denied the accusations. A second issue arose when Kuwait refused to allow Saddam a base from which to attack Iran during the war. Then there was the thorny issue of oil prices; Kuwait's unwillingness to decrease production to help raise the global price of oil was a major blow to Iraq. Furthermore, requests by Saddam for Kuwait to forgive the $14 billion in wartime loans they had provided to Iraq fell on deaf ears.

In addition to these issues, there was a territorial argument between both countries. Nestled between Saudi Arabia and Iraq, Kuwait seems an anomaly—at least politically. Iraq's western border with Saudi Arabia is relatively straight, moving from the border with Jordan in a southeasterly direction toward the Gulf. Roughly 100 kilometers from the Gulf, the border turns north and forms a semi-circle around the country of Kuwait (figure 2). This leaves Iraq with only a small appendage of land at its southern tip allowing access to the Gulf.

Kuwait was little more than a fishing village under control of the Portuguese in the 1500s, but soon became an independent sheikhdom and a major commercial center as trade between the Arabian Peninsula and India increased. With the Ottoman Empire to the north and the Persian Empire to the east, Kuwait was in a vulnerable position geographically. Britain decided the strategic value of Kuwait was worth the investment, and the small country became a British protectorate as part of the Anglo–Kuwait Agreement of 1899. After the discovery of oil in Kuwait during the mid-1930s, the British refused to give up control of the small sheikhdom, and it was only when the 1899 agreement with Britain expired in 1961 that Kuwait declared its independence.

By 1990, the British were long gone, and Saddam wanted Kuwait. Although he was unpopular both in the region and internationally because of his use of chemical weapons during the war, he was certain that other Arab states would acquiesce to his whims regarding Kuwait and that the West would simply not care. On July 25, 1990, the American Ambassador to Iraq, April Glaspie, told Saddam, "But we have no opinion on the Arab–Arab conflicts, like your border disagreement with Kuwait."[3] It all made sense—at least to Saddam—and none of his advisors dared risk the implications of expressing a dissenting opinion. On August 2, 1990, Iraqi forces invaded Kuwait. Within twenty-four hours they captured the country and set up a provisional government. By the end of the month, Kuwait was annexed and became Iraq's nineteenth governorate.

The international community was quick to condemn Iraq's actions. One day after the invasion, the UN Security Council unanimously passed Resolution 660, demanding that Iraq immediately and unconditionally withdraw all armed forces from Kuwait. More resolutions were to follow. On August 6, the Security Council agreed to a ban on all trade with Iraq, including oil and weapons. The UN then decided the annexation of Kuwait was illegal and imposed further economic sanctions. Saddam ignored the UN resolutions but was by then fully aware that invading Kuwait had been a grave mistake. Even his Arab partners turned against him.

The UN made one more attempt to intervene. The Security Council passed a resolution in November 1990 mandating that Iraq withdraw from Kuwait by January 15, 1991, and empowering states to use all necessary means to force Iraq out of the country after that date. There would be no conditions and no negotiations until Saddam left Kuwait. Nevertheless, Iraq was unwilling to leave without major concessions related to the repayment of wartime loans and control of the entire Rumaila oilfield. The situation was at an impasse. In the meantime, the U.S. and Britain attracted a coalition of thirty-nine countries to assist in removing Saddam from Kuwait. When Iraq failed to adhere to the January withdrawal deadline set by the UN Security Council, the coalition attacked in what became known as the Gulf War. It began with a six-week, constant aerial bombardment targeting both military and civilian infrastructure inside Iraq. The world was subjected to nightly displays of the bombing on TV news. More infrastructure was destroyed in those six weeks than in the entire eight years of the Iran–Iraq War.

Priority targets included anti-aircraft facilities, command and communication facilities, and military sites such as missile-launching sites and research facilities. Civilian infrastructure was also attacked. Ninety percent of Iraq's electrical production facilities were damaged, with attendant effects on water and sewage treatment plants that were dependent on electricity. Oil production facilities were also targeted. This focus on key infrastructure had major impacts on the civilian population of the country, particularly those living in Baghdad.

A report to the UN secretary-general from a UN mission to the region noted, "With the destruction of power plants, oil refineries, main oil storage facilities and water-related chemical plants, all electrically operated installations have ceased to function . . . untreated sewage has now to be dumped directly into the river—which is the source of water supply."[4]

Where did the untreated sewage that was dumped into the Tigris River end up? Downstream, in the Marshes.

The response from the Iraqi government was weak, but not insignificant, particularly for citizens of Israel and Saudi Arabia. Hoping that Israel would retaliate, which would likely bring other Arab countries to

Iraq's aid, Saddam fired missiles into the country. The physical impact on those living in Israel was minor, although there were injuries. The greatest fear was that Saddam would launch a chemical attack on Israel. While the rest of the world was glued to their TV sets, watching in amazement and horror at the devastation occurring in Iraq, Israelis were putting gas masks on their young children—and themselves—and running into shelters or hiding in their bathtubs.[5]

The ground campaign by the coalition forces began on February 24, 1991. Hundreds of tanks based in Saudi Arabia and western Kuwait attacked Iraqi occupying forces, while aerial bombing destroyed the main road from Kuwait to Basra, cutting off an important line of retreat. Thousands of Iraqi troops were killed. The road was later called the Highway of Death, both because of the number of fatalities and for the hundreds of military and civilian vehicles disabled and strewn about on the side of the road. As a parting shot to Kuwait, the retreating Iraqi troops set fire to over 700 Kuwaiti oil wells in what was called their "Scorched Earth" policy. It took almost eight months to extinguish the fires and cap the wells. One billion barrels of oil were lost—roughly 1 percent of Kuwait's total reserves—and global carbon dioxide emissions rose 2 percent because of the burning.

Four days later, the United States declared a ceasefire; Iraq had been driven out of Kuwait. Coalition forces moved to within 240 kilometers of Baghdad, but soon retreated to the Kuwaiti border. The UN mandate did not include regime change, and so Saddam survived the ignominious defeat at the hands of the coalition forces, although he quickly had to address the growing unrest within Iraq. The economy was in dreadful shape, owing to international sanctions, high inflation, and an urgent need to rebuild major infrastructure. The military sustained considerable losses, losing half their tank force; however, the elite Republican Guard force that had been created during the Iran–Iraq War survived after having been recalled to Baghdad early in the fighting. It was the regular army troops who took the brunt of the coalition invasion. And there were now civil protests in both the north and south against Saddam's regime.

Unlike the Iran–Iraq War, the Gulf War wasn't fought in the Marshes. There were impacts on the Marshes from the destruction of sewage treatment plants and industrial facilities, but these were short-term issues. More important was the role the Marshes played as a refuge for those fleeing the government. As Iraqi forces retreated from Kuwait, the Marshes provided a perfect place to hide, as they had done for centuries. Rather than being captured or branded as deserters and shot, many defected from the army and found refuge in the Marshes, much like troops who deserted in the war with Iran in the 1980s and the communists who hid from the government in the 1970s. Thus, they were away from the fighting and from Saddam's security forces. Saddam, however, knew that some of his army had melted away into the Marshes, and he vowed to address the issue when the opportunity arose.

On the evening of February 27, 1991, Jassim was having coffee with his father and other Bani Asad men in Hashim's mudhif. Reports from the BBC described uprisings in Basra after the U.S. and coalition forces removed the Iraqi army from Kuwait. Islamist groups, joined by army defectors and other Shi'a, took advantage of a perceived weakness in Saddam's regime and were now in control of Basra. The insurgency quickly spread to other cities in the south, including Amara, Hilla, Nasiriya, and Diwaniya.

One of the younger men in the mudhif excused himself and went outside to see whether there was any activity in town. He soon returned, affirming that people were marching in the streets. Jassim turned off the radio and began to hear noises and shooting coming from the center of town. As he left the mudhif, he saw a group of young people walking along the road connecting his father's island to the Chibayish city center, many shooting their rifles into the air. They called on the leaders of the tribes to join them. Jassim joined the marchers, eventually reaching the police station in Chibayish, which was controlled by rebel forces. All city offices were occupied by protestors, as was the main office of the Ba'th Party. Jassim was mildly surprised to find that some of the important leaders of the rebellion were not from Chibayish; he had never seen them before. The rebels later found five Ba'th Party officials hiding in their homes, and all were killed. The Shi'a Uprising had begun, and Chibayish was part of it.

In the aftermath of the Gulf War, Iraq's economic woes and devastated infrastructure made Saddam vulnerable to regime change, at least in the eyes of the Kurds in the north and the Shi'a in the south. Many were inspired by U.S. President George Bush's words encouraging Iraqi citizens to "take matters into their own hands and force Saddam . . . to step aside."[6] The U.S. seemed on board, and the Shi'a in the south responded.

Muhammad Baqir al-Hakim was a well-known Shi'a cleric who immigrated to Iran with thousands of other Shi'a after the execution of

Ayatollah Muhammad Baqir al-Sadr in early April 1980—not long before the outbreak of the Iran–Iraq War. Al-Hakim had been a major enemy of the Baʿth Party, and in exile he formed the Supreme Council for Islamic Revolution in Iraq (SCIRI). His goal was nothing less than overthrowing Saddam and establishing clerical rule. Saddam responded by having five members of al-Hakim's family executed, along with twelve other relatives. In response, al-Hakim created the Badr Brigade in 1982, the military wing of SCIRI. It was armed by Iran and staffed with Iraqi military personnel who had been exiled or defected.

The chaos following Iraq's withdrawal from Kuwait presented an opportunity for groups like SCIRI and al-Daʿwa to foment insurrection in southern Iraq. Sensing a weakness in the regime, the Shaʿban Intifada, or Shiʿa Uprising, began in late February 1991 when revolutionaries and defectors from the army took control of the Saad Square in Basra. The spark that ignited their revolutionary fervor was when a tank gunner fired a shell into a large mural of Saddam Hussein that dominated the square. Over the next few days, many other soldiers defected from the army and joined the rebels. Any portrait of Saddam was fair game. Although it seemed to be more of a spontaneous insurgency than an organized rebellion, SCIRI and al-Daʿwa provided both an organizational framework and an extensive network of supporters.

Within twenty-four hours, the revolution spread to all districts in the south. Thousands of people joined in, including Tahseen's seven brothers. Memory of their brother's execution gave them a strong motive to take part in the revolution. While they lacked Tahseen's religious fervor, they hated the Iraqi regime and were willing to fight against it. It seems everyone wanted to see the regime toppled. What or who would replace the government, however, was uncertain.

Saddam still had support from his loyal Republican Guard. He also had a considerable tank force that had survived the Gulf War. His response was merciless. He deployed chemical weapons and helicopter gunships, rounded up young men and shot them on the street, and had security forces conduct house to house searches. Unarmed civilians caught up in the fighting were indiscriminately killed. Many of the important Shiʿa holy sites in Karbala and Najaf were targeted and severely damaged in the fighting. It was a well-organized campaign of revenge against the Shiʿa. The rebels were ill-equipped and inexperienced, and resistance was weak.

Regaining control of the south took only a month, and as government troops regained the cities, there was a mass exodus of Shi‘a to the south, toward Iran and Kuwait. Residents used whatever transportation options were available to them: donkeys, trucks, boats, or even their own feet. Some tried to reach Kuwait and claim refugee status, while others hid in the Marshes and joined the army deserters. Anyone with close ties to Iran attempted to escape across the border. But Saddam's security forces were ruthless, and tens of thousands died.

Jassim, Suad, and their four children—another son, Dhafer, had been born a few months earlier—remained in Chibayish during the revolution. It was too dangerous to travel, and the threat of bombing by government troops was a real concern. Foreign radio stations reported on random killings occurring in Shi‘a cities in southern Iraq. An additional threat was from extremist groups participating in the uprising in Chibayish. Jassim was a government employee, and some believed him to be a man who had no religion and who supported the regime. And although most people in town knew Jassim and his family, the outsiders did not.

Few residents of Chibayish supported Saddam, but the reports of mass killings in other Shi‘a cities worried everyone, particularly after the Iraqi army blocked the single access road to Chibayish in both directions. Jassim and his family were no threat to either the regime or the rebel forces, and so they decided to stay in their home and hope for a peaceful resolution. But trouble searched them out. Jassim's main form of transportation to work was a small Fiat car that was owned by Mendes International, a Brazilian firm working closely with the provincial water ministry on the Main Outfall Drain. He had documentation proving that he was allowed to operate the vehicle but did not own it, nor was it owned by the governorate of Dhi Qar or the Iraqi Ministry of Irrigation. When staying in Chibayish, Jassim kept the car in his uncle's garage for safekeeping. Unfortunately, the rebel forces knew about the car and wanted it for their own use.

On the second day of the revolution, he was visiting his uncle's home when Fatima, Jassim's mother, came to see him. She told him that the rebels came to their home looking for him—they were taking possession of all government cars. Fatima pleaded with her son to give them his car. She asked for the keys, fearing that he might be shot if he refused.

Jassim was resolute: the car was not his and was not owned by the government.

"Don't worry," Jassim told her. "I'll take care of it. But you stay hidden here and don't go out until I get back."

At his uncle's garage were two police cars waiting for him and his uncle. Four men, their faces covered in black masks, were trying to figure out how to move the car. They explained they were taking possession of all government vehicles for use by the revolutionary forces and asked for the car keys.

"This is not a government car," Jassim responded. "Here are the papers."

The papers were printed in English and therefore incomprehensible to the rebel police officers. They handed back the papers and told him that unless he gave up the keys, they would take all three cars in the garage (Jassim's and his uncle's two cars).

Jassim still refused. "I want to see your director."

Jassim's uncle, who had been silent during this conversation, suddenly spoke up. He pointed to one of the policemen and said, "I know you. Your name is Majid and you worked for Saddam's security services. Now you are working for the revolutionary government?"

It was as much an accusation as a question. At that point, tensions reached a peak. The fellow he was pointing at refused to respond to the question and instead motioned for Jassim to get into the car. He told Jassim that he would take him to the director.

The two pulled up to the police station. Inside the main office sat Karim Naif, the administrative director of the revolutionary group controlling Chibayish, along with five senior sheikhs from the region. The sheikhs were there to declare loyalty to the new government.

The director looked past the sheikhs and then rose slightly from his chair. "Jassim!" he said in amazement.

Jassim was equally shocked. "Karim!"

The two had been good friends throughout high school. When Jassim enrolled in the University of Technology in Baghdad, Karim left to study petroleum management at the Oil Institute of Baghdad. They hadn't seen each other in years. Karim explained that he and others were trying to change the face of Iraq with the revolution. He then asked Jassim to sit down and talk to him, but Jassim was worried about his family. As much as he would have enjoyed talking to an old friend, he wanted to resolve the issue of the car and return to them.

"Okay, why did you come to see me?" asked Karim.

Jassim told him the story. He explained that his was not a government car but a private vehicle, owned by the company he worked for.

He revealed the documents, knowing that Karim could read English, unlike many of the others.

"Right now, you are not a government—only a revolution," said Jassim. "When you become a government, I will give you this car."

Karim seemed satisfied with the explanation and told Jassim he could keep the car. For now. If the revolution was successful, he would hold Jassim to his promise.

As Jassim exited the building, someone was yelling from the rooftop, ordering militia forces to various locations nearby. He walked back and forth on the roof with a noticeable limp.

"Who is that?" Jassim asked a passer-by.

The fellow told him it was Sayed Jabbar, the military leader of the revolutionary forces.

Jassim had never seen Jabbar before, although he had heard of him. Jabbar was not from Chibayish, and suddenly it was clear who oversaw the military activities in the revolution. It was not Karim, or any other Chibayish resident, but someone appointed by SCIRI.

Jassim returned to the garage and switched the license plate on his Peugeot with one from his uncle's car to make it appear like a local vehicle. He then moved the car to another location and covered it with reeds. A few of the rebels came looking for it the next day, but it was nowhere to be found.

The Iraqi army, supplemented by local police and heavy arms, tried to quell the rebellion in Chibayish in late March 1991. It was not an easy task. With only one road in and out of the town, rebel forces dug a trench in nearby al-Fuhood where they could remain hidden while watching army troops advance across the flat landscape. All the major cities in the south were back under control of the federal government, including Basra, Nasiriya, Najaf, and Karbala. Now the Iraqi army turned its sights on Chibayish—but first it had to capture al-Mawajid, a small village along the Euphrates River facing Hammar Marsh and just off the road between Chibayish and Nasiriya. Home to buffalo breeders, fishermen, and reed gatherers, it was a village that was rural and remote, and with a reputation for helping those in need.

The village extended along the Euphrates, mirroring the twists and turns of the river in a region known as al-'Abda'un. Until the 1980s, all the houses were constructed of reeds, their dried, light-brown color in stark contrast to the blue water and the green reeds surrounding the village. Today, the reed structures have been replaced by brick and concrete; traditional houses are a rare sight.

With many small villages in and around the Marshes, at first glance al-Mawajid hardly seemed noteworthy. Yet it was a village surrounded by dense reed forests and a puzzling array of narrow water channels and was a haven for those who were out of favor with the government. It had been so for centuries. This was well-known to the Ba'thist government, which during the 1970s planted security service agents in the region to help track down and assassinate key individuals who were deemed a threat to the state. Moreover, during the Iran–Iraq War and the subsequent invasion of Kuwait three years later, defectors from the military often found their way to al-Mawajid. So, too, did members of Islamic parties, such as SCIRI and al-Da'wa, who were hiding from the government. These groups helped form a major line of defense against the army as it moved to retake Chibayish.

The Iraqi army advanced on al-Mawajid cautiously. The flat landscape made them easy targets for rebels hiding in the reeds, and they were well acquainted with the history of fighting in the Marshes. Much to their surprise, however, the expected resistance from the residents of al-Mawajid never materialized. The army found the village empty. Everyone had scattered. Residents of al-Mawajid took enough food for three or four weeks and found refuge deep in the Hammar Marsh. Some eventually made their way to the highway linking Nasiriya and Basra, moving south until they were in coalition-occupied territory, near the border with Kuwait. In the town of Safwan they were met by representatives from the United States Agency for International Development (USAID) and applied for refugee status.

Chibayish now stood alone. Rebel forces put up a staunch defense of the town and fighting ground to a halt. The army commander wrote a letter to tribal leaders in Chibayish saying he would bomb the city if the rebels kept fighting. The tribal leaders held a meeting with Karim Naif, Sayed Jabbar, and other leaders of the rebel forces in the large mudhif of al-Kayoun, the former sheikh of all Bani Asad. Due to the apparent hopelessness of their situation, they decided to negotiate with the army to save the town. Led by Sheikh Tariq al-Kayoun, three emissaries from

Chibayish traveled to the village of Hammar, twenty kilometers to the west, to meet with the head of the army forces. After a short negotiation period, an agreement was signed between the two groups to end the fighting. Residents were given a ten-hour head start if they desired to leave before Iraqi government troops entered the city. The delay was just enough for the leaders of the revolution, along with their families, to disappear into the Marshes. Better to take their chances with the tantals than surrender and face the death penalty. Thus, a month after its inception, the Sha'ban Intifada was over. For some, however, the hardship was just beginning.

Leaders of the revolution were not the only ones to flee Chibayish prior to the arrival of the Iraqi army. Tahseen's father, Ali Kadhim al-Asadi—still one of the most respected people in town—was less sanguine about what might happen if he and his family remained in Chibayish. All his sons played a role in protesting and fighting against Saddam's rule, and Ali had little doubt what the repercussions would be if the army gained control. A few days prior to Iraqi troops entering Chibayish, Ali and his wife, along with five of his sons and two daughters, fled in small boats through the Central Marsh with six other families. Their goal was to reach the Iranian border and claim refugee status. Two other sons who had fought against the army, Salam and Wisam, remained out of sight in Baghdad amidst the chaos that had engulfed the city. Soon after Chibayish fell, however, both were arrested by security forces. Salam was sentenced to death, but this was later reduced to twenty years in jail. Wisam, who played a minor role in the revolt, spent four months in jail. It was years before the rest of the family found out what had happened to them.

Ali's family crossed the Tigris River and hid in the Hawizeh Marsh for almost three weeks as the Iraqi army regained control of the southern cities. Although the fighting ended in Basra and Amara (the cities to the south and north), their route to the east was blocked by Iraqi forces. Ali made a quick decision to head south, toward the border with Kuwait. The family packed into a rented mini-van and drove to the town of Safwan, between Basra and Kuwait. There they entered the demilitarized zone and were stopped by the U.S. Army and representatives from USAID. They explained what had happened, saying they feared for their lives and were being pursued by Saddam's forces. This came as no surprise to USAID representatives; thousands of Iraqis had preceded them.

Ali Kadhim and his family were placed in a camp in Safwan, along with 15,000 others. Among them was Adel al-Maajidy and his family from the village of al-Mawajid. A good friend of Jassim's uncle, Adel had escaped through Hammar Marsh with his entire village. Eight members of Adel's family paddled for three days through the Marshes and then walked for two days before finally reaching the Kuwaiti border. However, Kuwait did not want the refugees, and neither did Saudi Arabia. In addition, the coalition forces were preparing to leave Iraq, and although UN relief agencies and cease-fire observers would remain, they could offer no protection to the Iraqis in Safwan camp. A solution was required—and quickly.

Saudi Arabia eventually capitulated under international pressure and agreed to house the refugees for a short—albeit unspecified—time. U.S. forces then built a temporary camp in the northern Saudi desert. After two weeks in Safwan, Ali and his family were flown to the new Rafha refugee camp in northern Saudi Arabia, much like Adel al-Maajidy a few days before. The trip had been exhausting, but at least they were safe.

And yet Rafha proved to be a rude awakening for most of the Iraqi refugees. The average summer temperature in Rafha was well over forty degrees Celsius, with an average rainfall of zero. While most were relieved to be out of Iraq, their ordeal had just begun. The camp was a tent city in the middle of the desert. The tents would blow away in the wind, and sand covered everything, from dishes to people's faces. Most of the refugees had spent their lives surrounded by water; now there was only sand and fences.

Rafha housed over 33,000 Iraqis, most of them Shi'a Muslims from the Marshes and cities nearby. Saudi Arabia was not party to the 1951 UN Refugee Convention that set standards for refugee protection. As a result, conditions in the camp were deplorable. Fencing and barbed wire ringed the camp and no one was allowed outside the gate. The weather could be extreme, with scorching heat in summer, intermittent sandstorms, and frigid cold in winter. A nightly curfew was strictly enforced. Women had to be completely veiled and accompanied by a male family member whenever they moved around the camp. The UN and Saudi relief organizations provided food and medical assistance, but refugees were unable to work or leave. For many years, those in the camp were considered forgotten refugees by much of the international community.

Conditions in the camp improved slightly two months later, when the UN took control of camp operations. In the meantime, a more durable camp was being planned nearby. For the refugees, this meant only one thing: their stay in the desert would be longer than expected. Indeed, fourteen months into their ordeal, they moved to a second location, ten kilometers away. They were still in tents, but these were more durable and secure and somewhat larger. Ali's son, Ihsan, predicted that his background in civil engineering would allow him to work on construction and project management as the camp grew. However, this only lasted for two days, for while the Saudis were quite happy to have Iraqis work as laborers in cleaning, clearing waste, and even bricklaying, professional work had to be allocated to Saudi citizens.

People in the camps were an amalgamation of families, soldiers who had defected from the army after being defeated by the coalition, and single men who had fought in the Shi'a Uprising. Six months after moving to the second tented camp, Ali and his family relocated to a third new camp, this one designed for families and single women. There were cinder block buildings named simply "Block 1," "Block 2," and so on; at least these were equipped with kitchens and toilet facilities. Because of the size of Ali's family, they were given one of the larger blocks, a three-bedroom house. Yet although the camp felt more like a city, with street signs and shops, they were still in the desert. And two years into their ordeal, there was no end in sight.

The UN Convention Relating to the Status of Refugees was adopted in 1951 and has now been signed by 146 countries. It guarantees the rights and freedoms of refugees, including the right to not be forcibly returned to their country of origin. It also outlines the responsibilities of countries to accept refugees, although they are not required to do so. In general, the UN High Commission on Refugees (UNHCR) will refer people for resettlement, but the ultimate decision is made by individual countries. In the case of Iraqi refugees, their allocation to countries was a combination of selection and lottery. The U.S. agreed to admit no more than the total number of Iraqi refugees accepted by all other countries, whereas Australia had a strict requirement of a minimum high school education for men. Five years after leaving Chibayish, after interviews with both the UK and Australia, Ali Kadhim and his family were resettled in Sydney, Australia. Ihsan and his two brothers went first, and two years later they were able to bring the remainder of the family. Ihsan readily admitted that "it was nice to see water again."

Between 1986 and 2001, the Iraqi population in Australia rose from 4,500 to almost 29,000. Most of these were Shi'a or Kurds who fled the country after the 1991 uprisings.

"It took a while for us to understand the funny accents of Australians," Ihsan recalled. "My parents missed Chibayish and the Marshes, but we were very grateful to Australia for allowing us to settle there."

Two of Tahseen and Ihsan's brothers remained in Iraq. Salam, who was earlier sentenced to twenty years in prison, was released from jail after twelve years, just prior to the downfall of the regime in 2003. He returned to Chibayish and is now the director of administration at a health center. Wisam served his four months in jail and then moved to Basra; he presently works for the Directorate of Education in the city.

The fatigue and longing for family and friends took their toll on Ali Kadhim. He passed away in 2005, far from his homeland—but he had one more journey to take. In accordance with Ali's wishes, Ihsan and his mother returned with Ali's coffin to Iraq so he could be buried in the Valley of Peace in the holy city of Najaf. Their first stop was Chibayish, where the coffin lay in mourning for two days. Ali was well known from his days as principal of the primary school, and hundreds of people came out to pay their respects. He was then transported to his place of rest in the Valley of Peace.

Each year, Ihsan and his mother would return to Chibayish and Najaf to visit Ali's grave and pay their respects. They came back in 2020, as always; however, two years later, they were still living in Chibayish due to COVID-19 travel restrictions. The family's house where Tahseen and Ihsan grew up had been taken by the Ba'th Party after Tahseen's arrest in 1980 and converted to offices for Ba'th Party members. Ihsan's mother was finding the long trip back and forth to Australia exhausting, and so rather than return to Australia, she decided to remain in Chibayish, where she grew up.[7]

Adel al-Maajidy and his family shared a similar fate as most others in Rafha, although they did have one joyous occasion: their daughter Asma was born in the camp. After enduring four years of hardship, one of Adel's brothers was selected in the lottery to be resettled in the United States and the family was allowed to join him. Sponsored by Catholic Charities USA, they were resettled in Memphis, Tennessee. It was a welcome move, but a culture shock as well. Memphis might have been the home of the blues and the birthplace of rock music, but there were very few Arabic speakers in the city, and employment opportunities were minimal.

A year after arriving in Memphis, friends from Dearborn, Michigan, near Detroit, encouraged Adel to move north. Dearborn had the largest proportion of Arabic speakers in the U.S., and there were also job openings at the Ford auto plant. Almost six years after leaving al-Mawa-jid, Adel and his family arrived in a place they could call home. Another daughter, Assra, was born shortly after they arrived in Michigan. Today, the family still lives in Dearborn.

Prior to the al-Mawajid diaspora, few of the residents had gone to primary school, let alone university. Today, thirty years later, many speak fluent English and have degrees from well-known universities. Adel has since returned once to the Marshes to visit friends and family, but he has no desire to move back to Iraq. His daughters, Asma and Assra, have never visited Iraq, let alone the Marshes. Living in Michigan is a far cry from the tiny village of al-Mawajid. Few refugees returned to live there, particularly after the Iraqi government began its project to drain the Marshes. But as with the Marshes—and Jassim—al-Mawajid did not die.[8]

The brutal repression against the Kurds and the Shiʻa by the Iraqi govern-ment following the Shaʻban Intifada was received with almost universal condemnation. The total number of deaths and those who were disap-peared is unknown; however, a single mass grave found in the south three years after the uprising contained over ten thousand bodies. Thousands more Kurds and Shiʻa were displaced trying to escape the brutality of the Iraqi army and security services. The United Nations responded with Security Council Resolution 688 on April 5, 1991. The resolution did not criticize the government directly but condemned "the repression of the Iraqi civilian population" and demanded that Iraq "remove the threat to international peace and security in the region" caused by refugee flows.[9] Cuba, Yemen, and Zimbabwe dissented to the resolution as an intervention into the internal matters of the state. The dissent mattered little. Human rights issues were now at the forefront of discussions on international peace and security, and human relief efforts were part of this. To provide this relief and not be hassled by the Iraqi Air Force, the U.S., Britain, and France mandated a no-fly zone in northern Iraq.

Despite the UN resolution, the regime continued its fierce campaign of bombing local villages in the south. Saddam was not only targeting army deserters and Shiʻa rebels who escaped to the Marshes; he was tired of the general Shiʻa unrest and continued criticism of the Baʻth Party. He also worried about the strong ties many Shiʻa had with Iran. The Baʻth Party was convinced that Iran wanted to export its brand of religious fundamentalism to Iraq. Thus, punishing the Shiʻa for their actions wasn't enough; he wanted to destroy their livelihoods.

The U.S., Britain, and France were cognizant of the brutal repression of people living in the Marshes and the Shiʻa population in general. Saddam assumed these would be considered internal issues of state and that international forces would not interfere. The arresting and killing of Shiʻa rebels and their families by Iraqi troops and the indiscriminate bombing and burning of villages became international news; human rights groups put pressure on both the UN and key countries to respond. Using UN Resolution 688 as guidance, the U.S., Britain, and France acted independently and added a second no-fly zone to Iraq, this one incorporating the entire country south of Baghdad.

There was one loophole in the terms of the cease-fire agreement between Iraq and the U.S.-led coalition. Iraq requested that military helicopters be allowed to provide humanitarian support to populations in need, and the U.S. agreed. The helicopters proved very useful to Saddam, but humanitarian assistance was the last thing on his mind. Instead, they became assault vehicles against the Shiʻa, particularly those hiding out in the Marshes. Eventually, however, the threat of more sanctions on Iraq due to these activities limited Saddam's use of military helicopters, although it did nothing to affect the use of ground troops and artillery.

There was very little in the way of protection for residents, with attacks on the Marshes and the Shiʻa of the south continuing to be largely ignored by coalition forces. There was simply no international support for a continued ground war in Iraq by the international community. For one thing, Saddam had threatened to use chemical weapons if coalition forces began advancing on Baghdad. At the same time, most countries also expressed an unwillingness to support the growth of Islamic institutions with ties to Iran. Thus, the killing continued, as did the flow of refugees. But Saddam knew that fighting in the Marshes was challenging and time-consuming, and that eventually the U.S., Britain, and France would find a way to stop the massacre of civilians. He therefore began to look for alternatives.

Saddam and his advisors understood that continued military engagement in the Marshes—in the absence of attack helicopters—was fraught with difficulties. He also knew the history of the battles between Umma and Lagash, and Saladin's victories over the Crusaders in terms of the potential impact of cutting off water supplies to downstream residents or armies. Not only was he aware of the efficacy of such a strategy, but there was already a plan in place that would satisfy his needs.

In 1951, Frank Haigh, a British government engineer who was assigned to work with the Iraqi government, authored a report for the Iraqi Irrigation Development Commission titled "Report on the Control of the Rivers of Iraq and the Utilization of their Waters." The underlying principle of the report was that any water not captured for human use would be lost into the Marshes and eventually into the sea. It would, in effect, be worthless. Indeed, the perception that any water making its way to the Marshes would be wasted was a common one in the government in Baghdad, and in Syria and Turkey as well. Water only had value insofar as it could be captured for human use. The Haigh report proposed building a series of embankments and sluices, preventing water from reaching the Marshes. Much of the area could then be reclaimed for agricultural purposes.[10]

Saddam faced a dilemma after suppressing the Shi'a rebellion. He wanted to destroy the Marshes and the people living there but had to contend with both the no-fly zone and various foreign powers who were meddling in Iraq's affairs. Sitting on the shelf was the perfect solution: controlling water turned out to be Iraq's internal weapon of mass destruction.

The UN sanctions imposed on Iraq in 1991 after their invasion of Kuwait quickly took their toll on the Iraqi population. Inflation was rampant. One dinar could buy over three U.S. dollars in 1990; five years later, it took over 2,500 dinars to buy a single dollar. Prior to 1991, oil revenues allowed Iraqis to import much of their food. Now they were faced with minimal oil revenues and little food production capacity. Food rationing began in 1991. Families had limited access to staples such as sugar, rice, cooking oil, legumes, and flour. Disease and malnutrition increased throughout the country, with the urban poor being the most affected.

Jassim and his family were not immune to the impacts of the sanctions. There was strict food rationing, no internet access, no possibility of traveling abroad whether for business or leisure, and restricted communications with anyone outside the country. At a time of high inflation, salaries of government employees were slashed. Thus, most found it necessary to find ways to supplement their income to feed their families—Jassim among them. From 1995 to 2003, he used his brother's 1980 Nissan to shuttle people around Baghdad after work. If he saw someone standing by the side of the road looking as if they needed a ride, he would stop and offer his services. If people were surprised to be driven around by an irrigation engineer, they didn't show it; most needed a second or third job simply to pay for food and shelter.

The limited communication with those outside the country worked both ways. It was almost impossible for people from other countries to visit Iraq and describe the hardships endured by Iraqi citizens or the ongoing destruction of the natural environment. There were warning signs raised by UN relief agencies and a few non-governmental organizations, and at least one outside journalist braved the danger and sneaked into the country to report on the devastation,[11] but few were listening. UN Human Rights organizations tried to sound the alarm about the massive relocations and an increase in refugees from the south flowing into Iran, but no group or individual was able to document the tragedy well enough to convince the international community that action was warranted. The Iraqi government's use of chemical weapons during the Iran–Iraq War and their alleged pursuit of nuclear weapons dominated the international discussion.

People living in the Marshes were on their own. They had already suffered from the brutal suppression exercised by the government in quashing the Shi'a Uprising, and now they faced Saddam's imposition of an economic embargo against the south. And while fishing, hunting, and tending buffalo provided sustenance for those living in, or near, the Marshes, this situation was about to change.

In 1994, Jassim transferred from the Dhi Qar provincial government to the Iraqi Ministry of Irrigation in Baghdad. Not long after arriving at his new post, a document came across his desk from the minister broadly outlining a plan for future agricultural development in southern Iraq. It looked vaguely familiar to an old British report that assessed irrigation options for the region in the 1950s. Jassim looked through his files and dug up the original 1951 Haigh Report on controlling the rivers of Iraq. As soon as he read through the report, he realized the government's intentions: the agricultural development of southern Iraq meant little more than draining the Marshes. Jassim knew what the impact might be but was powerless to do anything about it. The government's plan was eminently and catastrophically successful.

Saddam's rage over the Shi'a in southern Iraq was not the only driving force behind his decision to drain the Marshes. Agricultural output in Iraq declined in the 1970s and 1980s, partly due to the war and partly due to the growth in oil revenues. The relative wealth of the state during that time allowed Iraq to import most of their food. Small landholdings were not economically viable, and the sector was controlled by wealthy landowners whom Saddam wanted to keep as allies. By the end of the war, the demand for locally produced food rose dramatically as oil revenues fell, and the large estate owners pressured Saddam to convert the Marshes to agriculture production. In turn, he used this to mollify the wealthy landowners, some of whom were part of his military elite. It was a win-win situation for the Ba'th Party: more food production and increased opportunity to provide inducements for support and allegiance.

There was also a concomitant need to expand oil production. Oil had always been a vital component of the Ba'thist modernization program, and there were major fields in the Marshes—particularly in southern Hawizeh, eastern Hammar Marsh, and along the Euphrates, in western Qurna. Draining the Marshes made drilling much easier, and oil companies were clamoring to build more wells and increase output.

A third reason behind the upstream withdrawal of water is that the government regarded it a waste to allow water to flow into the Marshes. Whether it was for hydro-electric production, drinking water, irrigation, waste management, or otherwise, water was an essential resource in the region, but for some, having it flow into the Marshes represented an economic loss to the country—better to divert it upstream for what was perceived to be more productive uses.

Finally, internal and external security threats made it essential for the government to have a modernized transportation system to move troops and heavy machinery if needed. Certain areas of the wetlands were an impediment to the rapid deployment of forces, and draining the Marshes was seen as a means of enhancing security.

Admittedly, these reasons were a side-show to Saddam's desire to exact revenge from deserters and rebellious elements who hid in the Marshes, to eliminate the livelihoods of Shi'a living in or near the Marshes, and to punish those who stood in the way of both modernization policies and Arab nationalism. Indeed, that Saddam had a grand scheme to drain the Marshes and kill or remove everyone—if not every living thing—is not mere hearsay. Documents show that a plan to rid the Marshes of undesirable elements was discussed and adopted by the Revolutionary Council as early as 1987. This included moving marsh villages to dry land and setting up an economic blockade of the Marshes that would include banning the sale of fish—a mainstay of the local economy. Saddam's initial solution was to bomb the area into submission,[12] but with a no-fly zone in effect, draining the Marshes and displacing the people seemed like a reasonable alternative—one that would likely not raise the ire of the international community, or at least not in the short term.

Implementing the Haigh plan was an enormous project and required coordination across a range of ministries. A military industrial council was created with the express purpose of removing water from the Marshes, thereby ensuring the forced displacement of the half million people whose lives depended on the Marshes. Saddam selected his cousin and son-in-law, Hussein Kamel, to lead the council. Kamel was a trusted aide who had created the much-feared Republican Guard during the Iran–Iraq War and organized an elite security unit around the president.

The council was comprised of ministers from agriculture, irrigation, oil, defense, and interior. A key position was the minister of interior, who was also in charge of national security. This position was held by Saddam's cousin (and Hussein Kamel's uncle), Ali Hassan al-Majid, better known to the west as "Chemical Ali." He was personally responsible for suppressing the Kurdish rebellion in the north in 1990, killing more than 100,000 Kurds in the process with poisonous gas or by execution. The council had every intention of repeating this process with the Shi'a of the south, albeit via more surreptitious means.

Constructing canals and embankments to divert water in southern Iraq began in early 1992. It was a relatively easy task, given that controlling water had been commonplace for over 4,000 years. Coordination from the various ministries expedited the engineering project, and it was only a matter of months before water in the Marshes began disappearing. It was as if a huge vacuum cleaner had sucked all the water out of the wetlands. The mass exodus of Marsh dwellers wasn't far behind. Satellite imagery provides striking visual evidence of the devastation wrought by the drainage activities. The total area of the Marshes inside Iraq in 1973 was 10,500 sq km, roughly the size of Lebanon (figure 18). This included a few smaller marshes not linked to the three main ones. The satellite photo is important because it provides a definitive baseline of the extent of the Marshes that can be used to observe and analyze future changes.

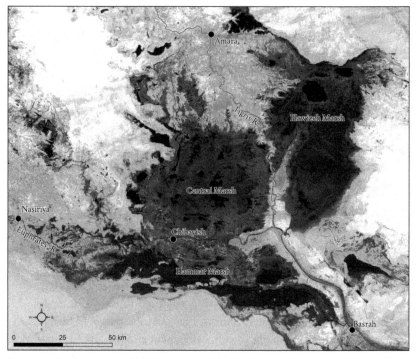

Figure 18. Satellite image of Marsh extent, 1973 (Landsat images courtesy of the U.S. Geological Survey).

By 2000, 90 percent of the wetlands had vanished (figure 19). Water remained in northern Hawizeh Marsh and in a few canals, including a wide strip just west of the Tigris River in the Central Marsh. However, except for northern Hawizeh, there was no longer any marsh. Hawizeh Marsh was at least partially immune from Saddam's transgressions because of water flowing from Iran; there was little Iraq could do to stop this flow, short of constructing a major dike along the border. This would happen fifteen years later, but the instigator would be Iran. The deep lake in Hawizeh, Umm al-Ni'aj, remained a prominent feature in the region.

One of the government's key projects to assist in draining the Marshes was named the Glory River or the Prosperity River. It appears as if someone placed a black ruler in the middle of the satellite image. The Glory River was a two-kilometer-wide, shallow canal just over fifty kilometers long that paralleled the Tigris River. Its sole purpose was to transport water away from the Marshes. The canal discharged into the Euphrates

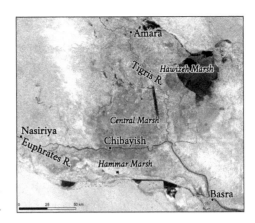

Figure 19. Satellite image of Marsh extent, 2001 (Landsat images courtesy of the U.S. Geological Survey).

River just to the west of its confluence with the Tigris. A smaller canal, the East–West Canal, looks like a piece of string dangling from the top of the ruler. This canal flowed thirty-five kilometers west to east before emptying into the Glory River, capturing any water flowing from streams to the north. These two canals, along with a series of locks and sluice gates[13] along the Tigris River south of Amara, were able to capture most of the water flowing from the Tigris into the Central Marsh, diverting it into the Euphrates and eventually out into the Gulf. The plan to drain the Marshes was so effective that the lower section of the Glory River appears dry in 2000; there simply wasn't much water left.

A second key component of Saddam's project to drain the Marshes involved the Main Outfall Drain (M.O.D.), often referred to as the Third River. By 1992, it was in its final stage of completion. Jassim had now worked for seven years on the M.O.D. after leaving the army, and while it was never intended to be used as a means of draining the Marshes for military or political purposes, it could easily be adapted to this purpose. Canals and diversions were then built to allow more water to enter the M.O.D. to be transported away from the Marshes. Jassim's work had the unintended effect of assisting Saddam's plan to drain the Marshes, and yet again he felt a tremendous sense of injustice. Two decades later, Jassim turned the tables and helped develop a plan to use the M.O.D. to reflood the Marshes, but for now it was simply another component of Saddam's grand scheme.

The third component of the drainage plan was to reroute the flow of the Euphrates River. Historically, when the Euphrates passed Nasiriya, it overflowed its banks and replenished the Hammar Marsh and the southern

Central Marsh. It was a relatively simple task for the government to build an earthen dam to divert the river into a separate canal known as the Mother of All Battles River, or Umm al-Ma'arik. In figure 19, it appears as if the river makes an abrupt right turn after Nasiriya and then flows south and east, around the Marshes. The Umm al-Ma'arik then merges with the M.O.D. southeast of Chibayish. Once the diversion was in place, the mighty Euphrates was little more than a trickle of polluted water, less than half a meter deep. As a result, no water could flow into the Hammar Marsh or the Central Marsh. In addition, the diversion effectively cut off the water supply to the town of Chibayish, where many Bani Asad lived, including Jassim and his family.

Construction of these two canals—the Glory River and the Mother of All Battles River—had a devastating impact on the Marshes. Any water that would potentially flow from the Tigris and Euphrates Rivers into the Marshes was diverted into the canals. It was like building two large pipes that would capture all the water before it could reach the Marshes, and then directing water through and around the wetlands. In the end, it drained an area roughly half the size of Lake Ontario. And yet construction took less than two years.

Living in constant fear of an Iraqi army that was destroying villages in the region and experiencing an increasingly dry marsh and a lack of fresh water for basic needs had a catastrophic effect on residents of Chibayish. Drinking water had to be delivered by truck. People who had access to a car drove either to Nasiriya or Qurna and loaded up with jugs of water. The fish were depleted, the buffalo had neither water to drink nor young reeds to eat, and many therefore perished, and the birds flew away in search of water. The population of the Chibayish declined from 63,000 in 1990 to 5,000 in less than a decade, and most left by 1995.

Throughout 1993, the army remained a visible presence in Chibayish, encouraging people to leave. Without water, food, or a means to earn an income, there was little choice. In September 1993, Jassim and his two brothers packed up the family and moved to a house that Jassim's brother Hazem had purchased in Hilla, a city of 350,000 in the Babylon Governorate and four hours' drive by car from Chibayish. Thirteen family members made the trip, including Jassim, his wife Suad, and their four children. However, they were not alone: during 1993 and 1994, two or three families moved from Chibayish every day. Jassim witnessed the town dying right before his eyes.

Figure 20. Buffalo returning home late in the day (courtesy of Mootaz Sami).

Muhammad, Jassim's father, hated living in the city. The family rented a store so that Muhammad might sell his leftover merchandise, but he was having none of it. The Marshes were special to him, and he couldn't stay away. Every two to three weeks, he would either cajole Jassim to take him on a work trip south or sneak out and rent a car and drive back to Chibayish, where he would stay with his brothers Ghani and Fadhel for three to five days at a time. Muhammad would return to his routine of having coffee every evening in Hashim's mudhif. Few of his friends remained, but his brothers did, and he missed them.

Despite the lack of water and the economic hardship experienced by the residents of Chibayish, some refused to leave. Abu Abbas, one of Jassim's uncles, was head of the al-'Awwad tribe and was committed to staying as long as there were other members of his tribe remaining in Chibayish. Others refused to leave their land, growing crops with the limited and often polluted water that remained. Some did not have the means to leave and preferred staying where they had lived their entire lives; they had no desire to become refugees in Iran and little interest in living in any city. They would rather die in the Marshes.

Jassim's father was never comfortable in Hilla. While he enjoyed being with his immediate family, his heart was in the Marshes. He knew the city was no place for him. Maybe the goblins and demons of the Marshes, the *'afarit* and jinn, knew it as well. In 2005, while walking across a busy street in Hilla, Muhammad was struck and killed by a truck. The Marshes could be heard moaning in sadness.

If there is a distinguishing feature of the Ma'dan—aside from their rather unique dwellings built from reeds—it would be the inescapable presence of water buffalo lounging in front of the houses like over-indulged royalty (which, in some sense, they are). These enormous beasts can weigh up to 900 kilograms, and when lying in front of a home they appear like colossal black or gray boulders guarding the entrance, with a snout and two curved horns emerging from the top—boulders that occasionally snort and groan.

When we speak of the Ma'dan (sing. Ma'di), we are talking about people living in the Marshes who own and tend to water buffalo. Residents in the Marshes who primarily raise crops, have cows, or fish for a living are referred to as Marsh dwellers. The distinction has become a bit muddled in recent years as Ma'dan have moved into towns—mainly to access better social services, but also because of poor water quality in the Marshes. They may live in town, but they keep their buffalo in the Marshes. Most Ma'dan own five, ten, or fifteen buffalo, although some control large herds that may exceed 150 buffalo. The animals are not mere ornaments for one's front yard; they are a vital source of milk, butter, cheese, and yoghurt—mainstays of the Ma'dan diet—and an important source of income. They can also sometimes be sold for their skins and meat. The buffalo also have another important role for the Ma'dan: their dried dung is an essential fuel for cooking. During the day, buffalo move, albeit very slowly and only with substantial encouragement, into the Marshes where they wallow in mud, feed on grasses, and drink water. In the evening, they return to their places of prominence next to—or sometimes inside—the house.

Sabbar was a Ma'di, and he and his family lived on a floating island of reeds at the northern end of the Hammar Marsh, south of the town of Chibayish. The island was roughly 500 meters square, built up over many years with Phragmites, the main species of plant in this part of the Marshes. The island doesn't actually float, but only appears to. The water is shallow, and layer upon layer of reeds are piled on top of one another—usually covered with woven reed mats—to construct a useable piece of land that rises above the water and is vulnerable to spring floods. The lives of the Ma'dan are inextricably linked to the buffalo and Phragmites. Young, tender reeds called hashish are used as fodder for the animals. Houses are also made of reeds, as well as the mats on which people sit and sleep. Sabbar's family sold buffalo milk and woven reed mats at the market, providing an income to help pay for marriages, clothes, cigarettes, and tea. Chickens and domesticated marsh birds, such as white storks, lived with the family as well. During the day, the small island was in constant animation.

Najmah, or "shining star," was born in 1976 on the family's island of reeds. She was child number twenty. Eventually, she would have eight brothers and fourteen sisters. Large families were the norm in the Marshes, and Najmah's father had three wives. The house Sabbar built for the family was much like those built in the Marshes 4,000 years before: it was designed as a large, open room that would not only hold his substantial family but their buffalo as well. Sabbar would often say that they were all his children, and he mourned their grief and took pleasure in their happiness. He was comforted by the activity around him, the vastness of the reed pastures, and the abundance of water.

Jassim's father and uncle also had a reed house in the Marshes, not far from Sabbar and young Najmah. In the late 1970s, Jassim was at university in Baghdad, but whenever he returned home to Chibayish he would spend time in his father's house. Most days, he would swim early in the morning in the canal in front of Najmah's home, sharing the water with children, buffalo, fish, and birds. Summers in the Marshes are extremely hot, and everyone takes refuge in the water, which is a cool respite for the children, but a necessity for the buffalo. Their large, dark bodies absorb much solar radiation, and they have an inefficient evaporative cooling system. For these beasts, water is for more than drinking; water means survival.

The water in the canal at the time was fresh, albeit a touch salty, with a salinity of roughly 300 parts per million (ppm).[14] However, fresh water does not mean clean water. Latrines were located at the edge of each island, four columns of reeds surrounded by a reed fence. Family members sat on two horizontal supporting branches as they balanced precariously one meter above the water and deposited intestinal waste accordingly. Although the waste provided food for the local fish population, it meant that dysentery, schistosomiasis, and other endemic diseases were rampant throughout the human population. Over time, however, most gained immunity, aided by the strong sunlight that helped fight disease.

Drinking water was obtained from adjacent canals that were free of human and large animal waste. Small pottery vessels helped filter, cool, and store water, making it refreshing to drink. This water was also used for cooking and bathing. The Ma'dan, like most Marsh dwellers, existed on a diet of fish, fowl, milk, and fresh cheese. Everyone was active and strong, and it gave their bodies, and their lives in general, a beautiful symmetry.

Even at a young age, Najmah tended to the water buffalo, giving feed to the young and collecting milk in metal containers to be sold by the men in the family every morning. Childhood is over quickly in the Marshes, and Najmah was soon part of the family's workforce. Her daily routine was to cinch her dress with a belt, push her two braids of hair behind her, grab a paddle, and maneuver her small chileakah forward. Najmah didn't have far to go, since the forests of tall green grasses emerged from the water less than 500 meters from her home. She would pull out a sharp, curved-blade sickle and cut the youngest shoots, storing them in the boat before returning home. She had gathered food for the buffalo.

Najmah enjoyed this work—being out in the vast space of the Marshes, with grasses that stretched seemingly to infinity, the freedom to select what area to harvest, and the joy of returning to her island house. She soon learned to fish as well, using traditional methods of fixing nets with dry, hard reeds. She placed the side with the hard reeds on the muddy bottom, which was never more than a meter-and-a-half deep, while the opposite side floated near the surface. Every few hours she would return to collect the fish that swam into the net.

Najmah entered primary school at the age of six, paddling her chileakah along the canal and across the Euphrates River, indifferent to the currents or the winter rain. The school was built in a semicircular shape, with arches of reeds and papyrus covered by a roof of woven reeds. Students sat on woven mats placed on the dirt floor; they were a mixture of boys and girls and had a male instructor. For opening day of school, Najmah wore shoes for the first time in her life and dressed in a new blue suit. She loved school and was a quick learner, and she soon knew the alphabet and a few short songs. After school, she paddled home to do her chores. Much to her dismay, however, when she reached the age of twelve her father no longer allowed her to attend school. The same was true for many of her female classmates.

Najmah was soon a young woman (having just turned fifteen), working for the family but thinking about the future. She was attracted to one boy, Gumar. One day, Gumar sang to her in a gentle voice:

Your cheek is like a pomegranate when it is opened,
I cried so hard my eyes no longer could see.
I am overwhelmed with your beauty and see only you.
And will not rest until I hug my beloved in my hands.[15]

Gumar was a slender boy with a melodious voice. The lyrics sparked her emotions and cheered her heart, and Najmah could not wait to see him every morning. When she went out to harvest the reeds, she would wander in her chileakah down narrow waterways until she met Gumar, in a private spot where they could be alone. One day she received her first kiss, and all she could think was what a beautiful world God had created.

Najmah had thoughts about marrying him, bearing many children, and living on their own small island. A joyous life. But the choice was not hers, and her thoughts and dreams were soon shattered. She would be married—not to the boy she loved, but to her cousin. It was as if a lightning bolt had struck Najmah, a sudden pain grabbing at her chest. She was overwhelmed with sadness. Still, the next morning, Najmah silently pushed her chileakah away from shore, in hopes of meeting Gumar. She was resolute: her heart could accommodate only the one she loved, and she was going to marry him. However, before the issue of marriage could be resolved, Najmah's life was again sent into turmoil.

The Iraqi government began to drain the Marshes and convert it to agriculture, giving the land to those who were loyal to the regime. For the people living in or near the Marshes, it was their worst nightmare. Najmah watched day after day as the water levels decreased, the salinity of the remaining water increased, and the areas of green pastures diminished. As the reed forests withered away and water levels dropped, the buffalo had nothing to eat or drink. The livelihoods of Marsh dwellers, based on buffalo, fishing, and reed harvesting, were in rapid decline, and the situation soon became dire for Najmah and her family. In early 1993, Sabbar decided to abandon their island near Chibayish and move his family out of the Marshes. They lost most of their buffalo: some died, while others were sold so the family could survive and save money for the dark days ahead.

Only three buffalo remained. The once beautiful environment was now almost completely dried out and destroyed, and Najmah, who was just shy of seventeen, was sadder than anyone.

The family packed up their belongings and, along with their remaining animals, brought them to Chibayish. It was a Saturday morning when a large, open-backed truck with two wooden slats on each side—just high enough to ensure nothing spilled out—drove down the main road in town and stopped next to the family. Najmah's father and brothers loaded up the truck with their animals, some worn furniture, and bags containing simple necessities of pottery, aluminum, and wood. Everyone climbed onto the rear of the truck except for Sabbar and the two youngest children, who rode in the front cabin. Najmah curled up in the back in shock and astonishment. She opened her eyes to peek at the deserted villages, demolished homes, dry canals, black-clad villagers, barefoot children, and the grim paved road ahead, writhing like a snake to nowhere.

Najmah's family was one of thousands who were displaced when the Marshes were drained. Families were leaving Chibayish every day, moving to the cities or other agricultural areas to find employment. It is estimated that up to 400,000 people were forcibly displaced, with almost a quarter of them going to refugee camps in Iran. Najmah knew that their story was not unique, but she still felt traumatized.

The truck passed through cities with houses made of bricks and concrete, and expansive agricultural lands interspersed with tall palm trees. Najmah longed for the water, the forests of reeds and papyrus, the birds, and the buffalo pastures. It was early evening when the truck turned onto a narrow dirt road just to the right of a trench full of reeds. Najmah saw a few emaciated buffalo—a stark contrast to the expansive agricultural fields laid out before her. She was thrown forward as the truck braked and came to a sudden stop. The family had reached their destination: a spacious vegetable farm, a cement block house, and a nearby canal. They quickly disembarked and removed their belongings. As soon as the truck was empty, the driver turned around and was gone. Life had changed in an instant.

When the sun came up the next day, Najmah and her family gathered around the breakfast table with the owner of the farm and a few members of his family. The owner welcomed the new arrivals and wished them a happy and productive stay on the farm, which was in the Mahmudiya District, thirty kilometers south of Baghdad and over 400 kilometers from

Chibayish. *This place will be our home for however long God wills*, thought Najmah. Sabbar's family were now agricultural laborers—they had little choice. The three buffalo that accompanied them would find food and water in the small canal and from crops that had been discarded.

Najmah was despondent. She had no idea what the future held or how she could communicate with Gumar, let alone see him. As the weeks dragged on, she felt as if she were carrying one of the buffalo on her shoulders. Still, she tried to settle into her new job and become comfortable with her new surroundings. She tended to the family's three buffalo, raised chickens, and planted okra, eggplant, corn, and cucumbers, all the while agonizing over not receiving news from her beloved.

One day in December 1993, a few months after their arrival in Mahmudiya, Najmah heard that an Ahwari family had migrated from the Marshes and settled not far from the farm where she worked. She begged her mother to allow her to visit. After much pleading, her mother acceded, on condition that she not stay too late. Najmah stepped confidently among the fruit trees, asking whoever she saw on her way about the new family of Ma'dan who had arrived the previous night. Finally, a passer-by guided her to where the new arrivals were staying. Her heart leapt in joy: it was Gumar and his family! She quickly ran back to tell her mother what she had discovered.

The cloud hanging over Najmah lifted in an instant. She still missed the Marshes, but at least she was able to see Gumar. The two families visited with each other often, sometimes over coffee and other times for dinner. She would meet Gumar under the shade of the trees when she accompanied the buffalo to the water trench at the edge of the farm. Gumar often hummed her a beautiful song and gave her kisses.

One evening, shortly after returning from seeing Gumar, she watched her father shudder in pain and fall to the floor sobbing, hitting himself on the face with both hands. She screamed for her mother and then rushed to help her father, asking what had happened. Her father admitted to her that Najmah's cousin Kamar, the boy she was supposed to marry, had been killed in a confrontation with security forces in the Central Marsh earlier that day. Later, as her father prepared to go to Chibayish to offer condolences to his brother's family, Najmah realized that the Marshes were no longer marshes. There was no water; they were nothing but dry desert, a miserable place where the government oppressed its people and shattered their dreams.

Najmah was torn. She understood her father's pain and anguish and felt sorry for her uncle and the family, but she never wanted to marry her cousin, and his death allowed her to openly have a relationship with Gumar. It would take time, however, since the family's grief would last for months.

"When pain crosses the boundaries of memory, the obstacle to marrying Gumar will be overcome, *inshallah*," she told her sister.

A year after the tragedy, Gumar's father approached Sabbar and requested that Najmah be allowed to marry his son. Sabbar refused, saying that the killing of his nephew was still too fresh in his mind. Najmah was distraught, but her mother and siblings pleaded with Sabbar to approve the marriage. In the end, he agreed, subject to waiting another six months. As was customary, everyone cited "al-Fatiha."[16] But the women did not sing with joy, respecting the grief of their family in Chibayish. As for Najmah, she was dancing on the tips of her toes, overcome with happiness.

Najmah and Gumar were married on the farm where Gumar's family worked as hired employees. They lived together in another non-descript, cement block house with Gumar's mother, father, and two sisters. Deeply in love, they were blessed with seven children over the next eight years. Despite these joys, neither felt free. They never became accustomed to a life where their well-being was controlled by owners of the land. The Marshes was still in their blood. It occupied a joyful place in their heart, a place where they had friends, and a place where they owned property.

Over the course of eleven years of work in Mahmudiya, few of the immigrants from the Central Marsh and West Hammar Marsh felt part of the social fabric of their adopted region. Their customs, beliefs, tales, and myths were different. Najmah, Gumar, and their families were part of a massive, forced migration of Marsh Arabs from their homes to places they had never been: Mahmudiya, Nasiriya, and Hilla on the Euphrates; Basra in the south; and even Iran.

In the spring of 2005, Najmah and Gumar received news that the Central Marsh north of Chibayish and the Hammar Marsh to the south were once again flooded and vegetation had returned. They were elated, and the decision to return was an easy one. Sabbar and his wife, along with Gumar's parents, had eventually found the transition to Mahmudiya a difficult one and passed away a few years before. Najmah and Gumar, with many of their siblings, rented two trucks and left behind farms in which they had no share or inheritance. Their relatives welcomed them back to

Figure 21. Marsh girl carrying dried reeds (courtesy of Meridel Rubenstein).

the Marshes with warm embraces. The day after they arrived back in Chibayish, the community responded and built a temporary reed hut for them until they chose a suitable place to build a home and raise their family.

A week after their return, Gumar and Najmah selected a place to live not far from the Euphrates. The choice was theirs: subject to agreement with the sheikh of their tribe, they could select any island that wasn't occupied, or share one of the larger islands with another family. It was a new experience for their seven children, who had never seen the Marshes. Gumar, with the help of other residents, built a house that would accommodate all of them. The following week, he bought a boat, a fishing net, a cow, and a sickle. It pleased them that the Marshes could provide them sustainable livelihoods in the form of fish, cane, milk, fresh herbs, and birds. It was all so comforting that they had two more children by 2009. Najmah's family—the one she had dreamed about—now consisted of eleven people and a host of animals.

Their eyes filled with joy seeing the water and vegetation and wildlife—the overall beauty of the place. Here the villagers pushed and paddled their boats through the Marshes to harvest the fresh green fodder; the reed huts embroidered the banks of the rivers and the multiple islands; and the new schools accommodated their children. They made new friends and came back to old ones. Najmah had struggled for over a decade to return to the Marshes. Finally, she was home.

5

GODS, GHOSTS, AND
THE DESCENT INTO CHAOS

Her house sinks down to death,
And her course leads to the shades.
All who go to her cannot return
And find again the paths of life.

(Proverbs 2:18–19)

The Epic of Gilgamesh is an early Sumerian poem that dates from the Third Dynasty of Ur (2100 BCE) in southern Mesopotamia. It is one of the earliest known pieces of literature in the world, and arguably one of the most important because of its apparent influence on the Bible, the Qur'an, Homer's *The Odyssey*, and even the myths of the Ma'dan. It is quite a remarkable story that embodies many of the themes of literature today. The poem includes heroes and monsters, a search for eternal life, love, sex, and betrayal, journeys to unknown lands, and, ultimately, questions about the meaning of life. Written on twelve tablets, the writing tells the story of Gilgamesh, the King of Uruk, who is two-thirds god and one-third human. He is a powerful ruler, but he is also abusive and disliked by the people of Uruk. Aruru, the goddess of creation, sends a wild man, Enkidu, to battle Gilgamesh. Enkidu is unsuccessful, but the two become close friends and soon embark on a journey together to the Cedar Forest, the sacred realm of the gods. There they battle and kill Humbaba, the demon who guards the sacred trees.

After their return to Uruk, Inanna (also known as Ishtar), the goddess of love, beauty, sex, war, justice, and political power, makes sexual advances toward Gilgamesh, but he rejects her. In retaliation, Inanna has her father

Anu, god of the sky, send the great Bull of Heaven to avenge Gilgamesh's rejection. The Bull doesn't have 40,000 legs, as in the Ma'dan story of the sapphire stone, but there are parallels. The beast brings drought and plague on Uruk, whereupon Gilgamesh and Enkidu slay the bull and throw its hindquarters in Inanna's face. This, however, turns out to be a big mistake and the gods decide to punish Enkidu for killing both Humbaba and the Bull of Heaven. Enkidu becomes ill and slowly descends into the dark underworld, where the dead wear feathers and eat clay. Gilgamesh is determined to avoid Enkidu's fate and embarks on a perilous journey to discover the secret of eternal life. He is led to a plant that grows on the bottom of the ocean, but when he places the plant on the beach, it is stolen by a serpent. The last half of the poem recounts his quest and its ultimate failure, and includes the story of the great flood. Gilgamesh eventually dies and the people of Uruk mourn the passing of a great leader.

In the Akkadian version of Gilgamesh, written roughly 1,000 years later, the twelfth tablet describes a Huluppu tree that grows in Inanna's garden in Uruk. The tree had been planted on the bank of the Euphrates River by Enki, Lord of the Earth. Enki was on a journey to a faraway land when the great South Wind ripped the tree from the earth and deposited it in the river. Inanna, who feared no man, pulled the tree out of the Euphrates and planted it in her garden. She cared for the tree and wanted to use its wood to build herself a throne. Ten years went by, and while the tree grew thick enough to harvest for construction, it also attracted some unsavory characters. These included a serpent that could not be charmed, and which wound itself around the roots of the Huluppu; an Anzu bird— half man and half demon—that breathed fire and water and made its nest in the top of the tree; and the dark and sinister Lilith, who built her home in the trunk of the tree.

Despite being disgraced at the hands of Gilgamesh, Inanna requested the king of Uruk's help in ridding the tree of the demons. And so he put on his armor, picked up his bronze ax, and set upon the creatures inhabiting the tree. He first slaughtered the serpent that could not be charmed. Then, he attacked the Anzu bird, who flew away to the mountains with her young. Lastly, he smashed the home of Lilith, who flew away to the wasteland. Gilgamesh then cut down the tree and carved a throne for Inanna.

The Epic of Gilgamesh provides the first glimpse of Lilith, the demonic figure inhabiting deserted places and appearing only at night. Lilith appears in many forms—and sometimes with different names—throughout Hebrew and Greek literature. She may come as a fairy of the night, sometimes friendly and other times evil and cruel. She may also appear as a beautiful maiden to deceive men. At other times, wild black hair covers her entire body and she aims to cause pestilence and misery. To the Ma'dan, Lilith is known as Salwa, and she lives in caves on islands in the Marshes.

Salwa is generally portrayed by the Ma'dan as having thick, fluffy black hair that covers most of her body except her chest. She has claws to kill her victims and tear off their flesh. Her eyes are red, her mouth is wide, and her teeth are long and sharp. At night, she is known to inhabit buckthorn trees, as many Ma'dan have discovered. There are multiple rooms in her cave dwellings, each with a different purpose. She spends most of her time in the great room, with her children. She can often be heard swirling around the borders of the Marshes, looking to kidnap men for marriage, whereupon she keeps them in small rooms of the cave, isolated from others. If one hears Salwa on a dark night, it is best to compliment her on her beauty and her cleanliness, whereupon she might leave one alone. Wearing an amulet around one's neck is also believed to help.

The Bedouin claim that Ali al-Swileh, a wise and brave man, boasted to the Khuza'il clan that he was not afraid of Salwa. No one believed him. One night, he ventured into the core of the Marshes, tricked Salwa into coming close, and then killed her with fire. He placed her dead remains in a bundle of reeds, brought it to Hamad al-Hammoud, the sheikh of the Khuza'il, and threw the bundle inside his mudhif. Al-Hammoud was never heard from again.

The buckthorn is a deciduous, flowering tree with elliptical leaves and thorny branches. It has yellow-green flowers and yellow or red fruit that can be quite sweet. The fruit is turned into jellies, jam, and vinegar, while the root bark of the tree is sometimes used for medicinal purposes as an anti-inflammatory agent. The buckthorn grows on the border of the Marshes—and it is said to contain evil spirits.

Passing through villages that lie adjacent to the Marshes, one will often see pieces of green cloth haphazardly hanging from the branches of a buckthorn tree. Green is a symbol of heaven and paradise; the Qur'an says that paradise is a place where people ". . . will wear green garments made of fine silk . . ." (18:31). It is also a symbol of nature and life, the color of the prophet Muhammad's cloak and turban, and the color most associated with Shi'a Islam. Green is a special color befitting the sacred buckthorn tree. Indeed, it is referred to as "al-'Alawiya" by the Ma'dan, meaning that it is believed to be an actual descendant of Imam Ali, the fourth caliph of Islam, and his wife Fatima al-Zahra'. Marsh dwellers never cut down the tree for fear that harm will befall them. The buckthorn also embodies great strength and metaphysical spirit; its leaves are used with camphor oil to bathe the dead before burial, in accordance with Islamic tradition. Unsurprisingly, therefore, the buckthorn is a central character in many of the stories, legends, and myths of the Ma'dan.

Jassim's grandfather, Muhammad Diwach, was out walking with friends late one evening when they overheard someone crushing coffee under a nearby buckthorn tree. The traditional method of preparing coffee for visitors to a mudhif is to crush the coffee beans with a metal bar in a copper container called a *hawwone*. It makes a very distinct sound that signals the coffee will soon be ready. When the boys went to investigate the sound, there was no one to be seen; it was as if the tree were laughing at them.

"The buckthorn is inhabited by *'afarit* and jinn," Diwach later told Jassim. "It is a sacred tree that often contains evil spirits."

Fahd al-Asadi was a friend of Jassim's and a preeminent storyteller. Born in 1939, he was considered one of the most prominent realist writers in Iraq, and his stories reflected the history, heritage, and affairs of local people. He passed away a few years ago in Baghdad, but his stories live on. When Fahd was a young boy, he lived next door to the manager of the small district of al-Midaina. In the manager's yard was a buckthorn tree, and Fahd would sometimes sneak over to his neighbor's yard late in the evening and pick ripe fruit off the tree. One day, the manager caught him taking the fruit and slapped Fahd in the face with the back of his hand, threatening him with more if he was caught stealing the fruit again.

When Fahd's father found out that he had been stealing the manager's fruit, he chastised him and warned him to stay away from the neighbor's tree. Upon Fahd's objection, his father sat him down and told him a story about the buckthorn. When Fahd's father was a boy, the storyteller in his village was named Shiheeb, whom they called Abu Mahdi. He worked as a cleaner in the health center and used to tell wonderful stories from *One*

Thousand and One Nights and other famous books to all the children. He also told them about the buckthorn. One night, as Abu Mahdi. approached the largest buckthorn in the village, he heard someone calling to him in a hoarse voice, but when he looked closer, there was no one there. He deduced that it must have been Salwa, the evil spirit. After hearing this, others in the village told of similar horrors involving buckthorn trees. People grew terrified, and the children were forbidden to listen to Abu Mahdi's stories from then on.

When Fahd told his friends about his experience with the buckthorn and his father's story, they didn't believe him. Instead, they started a competition to see who collected the most fruit from the manager's tree. Fahd was very fearful of being caught again but never mentioned it to his friends. The group of boys would place bets on who could sneak inside the manager's garden fence and steal the fruit. Who would be able to stand up to the evil spirit? Hammadi, a beautiful boy who used to roam through the orchards at night, was unafraid of the darkness and uncaring about the strange sounds emanating from the buckthorn trees. He soon became part of Fahd's group and told them stories that were confusing and scary. Fahd became terrified of the tree but could not resist; he wanted to see Salwa, the old woman, with her long molars, wild hair, and hoarse voice, who was often portrayed in stories by the villagers. Fahd used to see this old woman in his dreams, lying near the tree and wailing. People used to say that Salwa was part of the tree, and when a previous owner cut off a big branch, Salwa could be heard crying for her severed son. On that day, sap drained from the tree like it was shedding blood.

Many of the villagers told stories about their encounters with Salwa. One of them saw old Salwa riding on the back of the town butcher, Hamza, late in the evening, like he was a donkey. The butcher had stolen bricks from the manager's garden, right in front of the buckthorn tree. Umm Sabbar, another neighbor, told Fahd's mother that on the night of the memorial of the killing of Imam al-Abbas, she heard the sound of women moaning from the manager's garden. However, when she looked over the fence, she found no trace of the mourning women.

After that, Fahd wore amulets tight around his neck whenever he sneaked through the garden fence. He would read short passages from the Qur'an to drive away evil spirits before entering the yard—only then did he feel it was safe to approach the tree and steal the fruit. Indeed, the green pieces of cloth hanging from the buckthorn tree are not placed there simply to glorify that sacred tree; like Fahd's amulets, they are meant to keep the evil spirits at bay. Unfortunately, it doesn't always work.

The UN sanctions imposed on Iraq after its invasion of Kuwait in 1990 remained in force until 2003. The primary purpose—at least once Iraq withdrew from Kuwait—was to eliminate weapons of mass destruction that Saddam had used during the Iran–Iraq War and its aftermath, whether chemical, biological, or nuclear. It was also an attempt to diminish the military power of the regime and its role as a disruptor in the region. One of the requirements mandated that UN inspectors visit manufacturing sites in Iraq and assess whether the country had weapons of mass destruction and, if so, where they were stored. In addition, oil exports—the main source of revenue for the country—were severely limited under the sanctions, with roughly two-thirds of the revenues designated to meet humanitarian needs.

Few believed that Saddam could withstand the threat to his regime imposed by the sanctions; they expected that he would be forced not only to open his country to UN inspectors, but to relinquish power as well. After doing his best to impede their progress, Saddam allowed the UN Special Commission (UNSCOM) to access information and storage sites with regards to the alleged weapons of mass destruction. What was initially intended to be a four-month endeavor lasted eight years. By the end, UNSCOM concluded that the Iraqi regime no longer had weapons of mass destruction, nor the means to produce them. Nevertheless, despite economic hardships in the country and a weakening of the military, Saddam retained his iron grip on power through violence, coercion, and a purging of the military of any threats to his regime.

When Jassim transferred from the Dhi Qar regional government to the Iraqi Ministry of Irrigation in Baghdad in 1994, his family remained in Babylon. He would normally spend four or five nights a week in his office, sleeping in a separate small room that had a television, a stove, and a small refrigerator. A few other ministry staff had a similar arrangement, including the general director, senior administrators, and a few guards and official drivers. The schedule suited Jassim for both economic and professional reasons. Still, although his new job brought more responsibility, it coincided with a severe down-turn in the economy. Salaries for government employees decreased relative to the U.S. dollar due to the drop in the

Iraqi dinar, food was rationed, and travel to professional meetings banned. There was no internet and only limited communication with people outside the country.

Rapid inflation and the relative unavailability of food and medicine resulted in the UN instituting an oil-for-food program in March 1997. A year later, when the magnitude of the problem became clear, oil revenues were allowed to treble, with two-thirds still going to humanitarian assistance. Businesses involved in the illegal shipment of goods prospered. And yet nobody in the country dared criticize Saddam. In 1998, Madeleine Albright, U.S. Secretary of State, announced that UN sanctions would not end until Saddam had been deposed. The official policy of the U.S. was now that of regime change in Iraq.

The bombing of the World Trade Center and the Pentagon in the U.S. on September 11, 2001 brought increased attention to Iraq. Although no link between Iraq and the attack on the U.S. was ever found, President Bush and his advisors believed one existed and set out to convince the American public to support further action on Iraq. Plans for an invasion of the country were drawn up by the U.S. in the fall of 2002, and efforts commenced to convince their NATO allies to join in a coalition to implement the plan. While the British government was supportive, France and Germany equivocated. Instead, they pushed the UN Security Council to authorize yet another weapons inspection of Iraq. It mattered little to the U.S. that the previous UN assessment found no evidence that Saddam had weapons of mass destruction; the decision to invade had been made regardless.

Opposition to an invasion was widespread. Canada, France, Germany, and Russia, among others, refused to support a war with Iraq. Without their support, there was no UN resolution, as there had been after Saddam's earlier foray into Kuwait. Regardless, the U.S. administration was convinced that Saddam still had weapons of mass destruction and that the regime supported terrorist groups, and they vowed to do something about it. The U.K., Australia, and Poland were willing to join the U.S.-led coalition. Together, they invaded Iraq on March 12, 2003.

Unlike in 1991, when the Iraqi army was routed and pushed out of Kuwait, this time the coalition forces met strong resistance, particularly in cities like Basra, Nasiriya, and Karbala. After three weeks of intense fighting, they entered Baghdad, quickly gaining control over the airport and moving troops into the southern part of the city. The initial invasion phase lasted from late March until May 1, 2003, when U. S. President Bush formally declared an end to combat operations and announced that a provisional government would be assuming power in Iraq. It took a few more months of searching, but on December 13, 2003, Saddam was captured at a farm in al-Dawr, near Tikrit, 200 kilometers north of Baghdad. Three years later, he was convicted of crimes against humanity and executed by hanging on December 30, 2006.

When coalition forces entered Baghdad, they quickly gained control of the airport and moved their troops into the southern part of the city. Jassim found himself trapped in his office in the Ministry of Irrigation building in the al-Karadha District of Baghdad. The intense fighting between Iraqi troops and the invading forces meant it was too dangerous for him to return to Babylon to be with his family.

The only items of value to Jassim in his office were his books. He had nearly 200 of them, on topics ranging from poetry and philosophy to engineering. Fearing that his office building might be damaged in the invasion, he moved all his books to a friend's house for safekeeping. As the battles ensued between Iraqi forces and coalition troops, foreign television was the only source of unbiased information for Iraqi citizens on the invasion—although accessing non-Iraqi channels via satellite dish was illegal. Radio, newspapers, and the one television station were tightly controlled by the regime. There was no freedom of the press in Iraq, and any independent voice was quickly silenced.

Jassim was fortunate. The Ministry had a satellite antenna, so he could access an Iranian television station, al-'Alam TV, from Tehran. This provided him with a more accurate perspective on the invasion. And while he learned about the battles from TV news, he could also hear bombs exploding as they destroyed key buildings and vital infrastructure in Baghdad. Coalition forces led by the U.S. entered the city on April 9, 2003, and by April 12, the fighting was over. The residents of Baghdad flooded downtown streets and squares and began celebrating as soon as they learned coalition forces had entered the city. After a quarter century of Saddam's repressive actions and over a decade of crippling economic conditions, people wanted retribution. Government buildings were among their first targets. Jassim watched the events unfold on television and learned about rioters entering government buildings and taking what they could. It was time to leave.

Jassim fled the ministry building on the afternoon of April 12, just before a crowd entered and began going from room to room, stealing anything of value and destroying the rest. He felt that being a government employee might lead some to believe he supported the regime, which was certainly not the case. Deciding it best to leave the office and return home to Babylon by whatever means possible, Jassim walked quickly toward the National Theatre, where a large crowd was celebrating. As he made his way through the crowd, he heard a voice behind him shouting his name. He turned to see a friend and retired colleague from work, Sayed Abdul Zahra, approaching him. His friend asked him why he was wasn't with his family in Babylon, to which Jassim explained that he had been staying in the office, but when he looked out the window and saw a large crowd of people moving toward the building, he ran down the stairs and out through a side door.

"And what are you doing here?" Jassim asked Abdul Zahra.

Abdul Zahra told him there was an unexploded missile sticking out his back garden. He immediately gathered his wife and daughters and moved them to a hotel in the city for a few nights. It was late in the afternoon, and he told Jassim they had ample room in their suite and invited him to stay with them. Jassim knew it would be dangerous to try and leave Baghdad that evening, particularly since he had no means of transportation. He thanked his friend and agreed to stay for one night. As Abdul began walking toward the hotel, Jassim stopped him and asked his friend to drive him back to the ministry to retrieve a bag that he had left behind in his office.

Abdul Zahra looked incredulous—there were looters everywhere. But Jassim was adamant. He explained that he left some important stories and poems he had written in a cloth bag, which he stashed beneath a portable closet that was bolted to the wall. In his haste to exit the building, he had forgotten all about the bag. The poems and stories were important enough to Jassim that he was willing to fight through an unruly crowd to retrieve them. His friend reluctantly agreed.

The two made their way to Abdul Zahra's hotel, found his vehicle, and drove to the Ministry of Irrigation building. The streets were clogged with people, and traffic was barely moving. As they approached the ministry, Jassim looked up at his third-floor office and all he could see were smashed windows and papers floating in the breeze. There was debris scattered all around the building, but most of the looters appeared to have moved on. The safest way to enter the building was through the basement garage. While his friend waited in the car, Jassim made his way to the side entrance, slowly opened the door to the garage, and peered inside. Two men and a woman were stealing new tires from a once-locked storage container in the basement, but it was otherwise quiet. They glanced briefly at him and then carried on with their pillaging.

Jassim climbed the back stairway three stories to the office. The furniture, appliances, and computers were missing, and glass and papers were scattered everywhere. Miraculously, the portable closet remained intact, with the bag still in place underneath. Jassim grabbed the bag, exited his office, and raced down the stairway back to the garage. This time, however, the thieves took notice. One of them walked toward Jassim and in a menacing tone asked him what was in his bag.

Jassim believed the man thought he was sneaking money out of the building. "Just my personal papers," he responded. "Here, I'll show you. Only some poems and stories I wrote."

He emptied the contents of the bag onto the garage floor. It was clear there was nothing of value in the bag, and the thief returned to his task of removing the tires. Jassim gathered his papers, stuffed them into his bag, and exited the building. Abdul Zahra was waiting anxiously in the car, and they quickly drove away. There was too much chaos in the streets to hang around.

The next morning, Jassim thanked his friend and his family and set out on foot toward Babylon, carrying his bag of poems and stories. He first went to Firdos Square, to see where the statue of Saddam had been pulled down by Iraqis three days before, with help from the U.S. military. He then walked toward the one remaining bridge over the Tigris River and soon came to a military checkpoint. A U.S. army officer who appeared to be in charge watched him approach. He stepped in front of Jassim and asked him where he was going.

Jassim handed over his papers that confirmed he was an employee of the Ministry of Irrigation and told the officer he was on his way home to Babylon to see his family.

"On foot?" asked the officer.

"If I have to," responded Jassim.

He looked through Jassim's papers and waved him on.

Jassim thanked the officer and then asked, "If you stop a car going my direction, could you request that they pick me up and give me a lift?"

Very few cars were on the road at this point and the officer seemed dubious. Just as Jassim was leaving, however, a mid-sized, black sedan pulled up to the checkpoint. The driver was a middle-aged man with black hair and a moustache. An elderly woman sat in the passenger seat, and two young women were in the back. The driver explained to the officer that they were trying to reach their home across the Tigris. The officer nodded, and then looked at Jassim. He turned back to the driver and motioned for him to open the back door so Jassim could join them, ordering the driver to give their new passenger a ride as far as they were going.

Jassim was surprised, but no more so than the women in the car. When the driver opened the back door, Jassim slid into the seat next to the two young women. They wore outfits that looked to him like pajamas, and their heads were uncovered. The older woman in the front seat, however, wore a traditional robe, and her head was covered with a hijab.

All were quiet for a few minutes while the U.S. officer checked their papers. The security gate finally opened, and as the car moved slowly over the bridge, one of the young women leaned forward to the woman in the front seat. "Mom, can you give me a cigarette?"

The mother opened her purse, pulled out a cigarette, and handed it to her daughter.

She next turned to Jassim. "Do you have a light?"

"No, sorry, I don't smoke," he replied.

The driver lifted his head slightly and glanced in the rearview mirror. "Are you Iraqi?"

"Yes. I'm an engineer who works at the Ministry of Irrigation and I'm trying to get home to Babylon to see my family."

The women and their driver noticeably relaxed on hearing this. They had assumed Jassim worked for the U.S. Army, possibly as an informer.

The driver passed the older daughter a lighter, and she proceeded to light her cigarette and take a few puffs. Then she turned to Jassim and began telling their story. She was twenty-one years old and married with two children. Her sister was only sixteen. They had just been released from the women's prison in Khan Bani Saad, a village in Baquba Governorate, fifty kilometers northeast of Baghdad.

When coalition forces entered Baghdad, the guards opened all the jail cell doors and left the prison; the prisoners were free to go. When the two young women walked out of the building, they saw their mother who'd found out her daughters were being held in Khan Bani Saad and come looking for them. The two young women had spent four months in jail, and Jassim now understood that their shabby clothes were standard issue at Khan Bani Saad. The family was on their way back to the mother's house to be reunited with the older sister's children.

After taking a few more puffs on her cigarette, the woman resumed her story. Her husband previously worked for Uday Hussein, Saddam's older son. Uday thought her husband was stealing from him and sent his thugs to their house to arrest him. The husband, however, was forewarned and escaped to Syria. When Uday learned of this, he came back and arrested the two sisters and threw them in jail. Initially, they were taken to Uday's private jail in the building of the Iraq Olympic Committee but were eventually transferred to the al-Habibiya women's prison in Baghdad before being sent to Khan Bani Saad.

Uday Hussein, thirty-nine years old, and his younger brother Qusay, would be killed in Mosul during a battle with U.S. troops three months later. Uday was, by all accounts, more ruthless than his father, often to both friends and enemies. He killed with impunity. Saddam appointed Uday chair of the Iraq Olympic Committee and the Iraq Football Association in 1984. As head of these organizations, Uday was renowned for jailing and torturing athletes and coaches who failed to win medals or score goals. He would place them in his private prison inside the Olympic Committee building and sometimes let them out to participate in events or games. Many athletes left the country rather than deal with Saddam's son.

Uday also founded and commanded the Fedayeen Saddam, a volunteer paramilitary force of 30,000 to 40,000 members who answered directly to him rather than to the military establishment. They were among the fiercest fighters against the coalition forces in 2003. Survivors of Fedayeen Saddam became part of the Iraqi insurgency that battled the government from 2003 to 2011. Uday was Saddam's heir apparent until an assassination attempt by a small Islamist resistance group left him unable to perform his duties. He was shot seventeen times, and although he survived, Uday sustained severe nerve damage and lost the use of one leg. The torch then passed to Qusay. Still, to say that the two young women sitting next to Jassim in the car were frightened of Uday was an enormous understatement. Everyone was frightened of Uday.

Jassim could relate to their stories of incarceration, but he had little desire to dwell on the past. He was anxious to return home. The women and their driver dropped him off on the main road before reaching their home and wished him well. He started walking. The road from Baghdad to Babylon had been closed to everything but military traffic for the previous ten days. When Jassim began his journey, he was among hundreds of others who were walking on the side of the road, attempting to return home after being stuck in Baghdad.

U.S. military vehicles were traveling in both directions, but few personal vehicles were allowed. After seven or eight hours and roughly forty kilometers of walking, Jassim came upon a long-distance taxi service that took passengers to specific cities, including Nasiriya, Najaf, and Hilla. He chose the one for Hilla and was home by late evening. One of his brothers had managed to avoid coalition troops by driving a back route to Baghdad searching for him, but in vain. After two days of looking, he returned to Babylon to find Jassim already home.

Saddam Hussein's regime had been defeated and there was no longer an Iraqi government. The U.S. and its coalition partners established a temporary government to fill the void until democratic elections could be held. What was called the Coalition Provisional Authority (CPA) was now in charge at the federal level and in all eighteen governorates. The CPA was led by American and British personnel. Iraq was now occupied by foreign powers for the first time since independence.

Only one government building in Baghdad had not been looted or vandalized: that of the Ministry of Oil. U.S. troops secured the building on their arrival in the city, and it was heavily guarded. The large, six-story building was in Ziyouna District, not far from Tahrir Square. The square is well-known for being home to the Nasb al-Hurriyah monument celebrating key events leading to the establishment of the Republic of Iraq in 1958. It was also the site of anti-government protests in 2019. The Ministry of Irrigation, which had been re-named the Ministry of Water Resources (MoWR) by the CPA, was assigned the sixth floor of the building.

Jassim returned to work after spending two weeks with his family in Babylon. As before, he took a long-distance taxi since his work vehicle was still in Baghdad. Jassim was now in a temporary office and had a new supervisor. Dr. Hassan al-Janabi, an Iraqi who had been working in Australia, returned to help the coalition government restructure the MoWR.

Figure 22. Young girl and buffalo (courtesy of Mootaz Sami).

The plight of the Marshes was now international news, and al-Janabi soon established a separate office within the ministry named the Center for the Restoration of the Iraqi Marshes (CRIM). Jassim became CRIM's senior technical engineer. One of his first assignments was to compile a comprehensive report on the Marshes, with recommendations for future actions. Not even Jassim was prepared for the devastation he would discover.

There is little doubt about which is the most famous animal ever to escape from the Marshes. Despite the existence of lions, wild boars, wolves, poisonous snakes, a multitude of beautiful birds, and, of course, the ubiquitous water buffalo, the only live animal to make the journey from the Marshes to London (and eventually Scotland) was Mijbil (or Mij), the smooth-coated otter. His travels and antics, along with his unfortunate death at the hands of Big Angus, were etched into the minds of millions of British and North American readers when they read Gavin Maxwell's 1960 book *The Ring of Bright Water*.

Smooth-coated otters are reddish-brown animals, with short, sleek fur and a flat tail. They have a slightly rounded head, legs that are stumpy and strong, sharp claws, and can weigh as much as eleven kilograms.

Smooth-coated otters live in holts or dens along the banks of rivers and canals, generally under tree roots or amidst boulders. Their diet consists mainly of fish, although they will hunt rats and even eat snakes at times. In the wild, smooth-coated otters live and hunt in groups of up to eleven. And although they are commonly found in the swamps and rice paddies of India and Southeast Asia, one subspecies of smooth-coated otter—named Maxwell's otter, after Mij's owner—only exists in the Marshes of southern Iraq.

A second species of otter, the Eurasian otter, is also found in Iraq, in both the Marshes and lakes in the far north. The Eurasian and smooth-coated otters are friendly creatures that, when provoked, might also bite your fingers off. Nevertheless, aside from being endearing and fascinating, Otters play an important ecological role. They are an indicator species, meaning that their presence is a sign of overall ecosystem health. When the number of otters increases, it is a sign that water quality is improving, and fish are abundant. In addition, some otters are critical in maintaining ecosystem balance, acting as predators for invasive species lower on the food chain.

The Eurasian otter was once widespread in the Middle East, but its numbers are declining. The International Union for Conservation of Nature (IUCN) rates the Eurasian otter as near-threatened and needing conservation measures to survive. Drift nets used in fishing are a major cause of their demise via strangulation and drowning, along with hydro-electric projects, the draining of wetlands for agriculture, poaching, and pesticides.[1]

Gavin Maxwell was a naturalist from Scotland who visited the Marshes in 1956 at the invitation of Wilfred Thesiger. He toured the Central Marsh and, after seeing otters playing in the marsh, decided to take one back to Scotland as a pet. Maxwell, with Thesiger's help, selected a suitable young otter and named him Mijbil (or Mij), after a sheikh he met in the Marshes. It did not end well for Mij. Less than a year after leaving the Marshes, Mij sneaked out of the house—likely searching for Maxwell—and crossed a nearby road. There he ran into Big Angus, a road worker, who mistook him for a large rodent and killed him with a pickaxe.

Mij was not the only smooth-coated otter from the Marshes to suffer an early death. Hunting devastated the population, and in the 1970s the IUCN labeled smooth-coated otters a vulnerable species. After the Iraqi government drained the Marshes in the early 1990s, the otters were believed to be extinct. The initial report by UNEP in 2001, highlighting the extent of damage to the Marshes, noted that destruction of the wetlands would almost certainly lead to the global extinction of the endemic smooth-coated otter sub-species.

Otters are wily creatures, however, and find ways to survive that are sometimes beyond human comprehension. The skin of a Maxwell Otter was found in the Amara market in 2008; it was from Umm al-Ni'aj, the deep lake in Hawizeh Marsh, and had been killed by electrocution.

Indeed, two surveys of smooth-coated otters conducted in 2012 and 2013 confirmed that there were at least a few Maxwell Otters remaining in the Central Marsh as well. It was a revelation to the nature conservation community inside Iraq and around the world. Somehow, Mij's relatives had survived, demonstrating the remarkable resilience of wetland ecosystems, even in the face of disasters—assuming the ecosystem is not permanently damaged.

Water continued to flow into the Marshes after the ad hoc breaching of embankments in 2003–2004. Soon levels approached those of two decades before—at least in areas that had been reflooded— and in 2005, people began moving back to the Marshes and villages on the edge of the wetlands. Prior to the draining, the three major marshes exhibited a similar mix of animals and plants. The northern section of Hawizeh Marsh was the only area to survive unscathed the ravages of being completely drained, and thus provided a useful comparison to demonstrate the resilience of the other marshes.

Reeds and grasses appeared not long after water reentered the dry marshes. Aquatic plants and fish accompanied the water that flowed from adjacent canals once the embankments were destroyed. Bird life soon followed. It appeared that the Marshes might be more robust than expected. The unique way of life of the Ma'dan and other Marsh dwellers, the life that Jassim knew as a young boy, might be preserved after all.

Much had changed, however. Plants normally found in brackish water, such as the pond weed, appeared. Invasive species such as whorled water milfoil and torpedo grass arrived and made it difficult for traditional species, such as floating fern and duckweed, to survive. These changes indicated that water quality was lower than before. In fact, the water returning to the Marshes was loaded with salts and other pollutants, and therefore undrinkable. Agricultural runoff and saline sediments polluted the canals and, in turn, entered the Marshes. Added to this were large amounts of untreated wastewater, the result of infrastructure that had been damaged by coalition forces during the Gulf War and never repaired. Municipal wastewater was now discharged directly into the rivers, eventually entering the Marshes. Phragmites, the essential reed for building and buffalo fodder, returned to the Central and Hammar Marshes, as did hornwort, but the plant ecology in these two marshes was noticeably different, as was their overall health.[2]

The wetlands of southern Iraq are an important wintering ground for migratory birds. It is uncertain whether the migratory and resident birds normally found in the Central and Hammar Marshes moved to Hawizeh during the twelve years after the draining, or simply found homes and wintering sites elsewhere. But when the Marshes were reflooded, they did return. Egrets and gulls continued to be the most common birds throughout the Marshes. And although there were some differences—with the little tern now populating Hammar Marsh, and two species of heron, the squacco heron and the purple heron, becoming more prevalent in the Central Marsh than elsewhere—by 2005, the distribution of bird species in the two marshes was comparable to that of Hawizeh. A few species, however, fared less well.

The Basra Reed Warbler is an endemic bird with a long, pointed bill that once thrived in the dense reeds of the Marshes. By 2003, it was considered an endangered species due to habitat loss that began in the 1970s. The warbler was seen in small numbers in all three marshes after 2005, but remains vulnerable to the continued cycle of wet and dry years. The Marbled Duck, which is also resident in the Marshes, and the White-Headed Duck, a wintering bird, remain on IUCN's vulnerable list and are very dependent on the availability of marsh habitat. Nevertheless, their numbers have risen in recent years.

Sixty percent of the fish consumed in Iraq prior to the draining came from the Marshes. The most profitable species for local fishermen were two species of barbels, the shabout, and the bunni. As early as the 1970s, the Crucian Carp, a species brought in from Iran, was able to thrive in a range of conditions and became a commercial success. It also began to outcompete native species, such as the shabout and the large gattan—ideal fish for cooking on an open flame.

Stagnant water, which had become common in the Marshes due to a lack of flood water, often becomes eutrophic. When this happens, the concentration of nutrients like calcium and magnesium increases which, in turn, promotes algae growth. The additional plants consume oxygen that fish need to survive. As a result, the bunni and carp—species that could withstand lower levels of oxygen—became the dominant species in the Marshes. The Central and Hammar Marshes also included other species of fish that could withstand higher levels of salinity, such as mullet and mossul bleak. Brine shrimp, a species not found in fresh water, were seen in the Hammar and Central Marsh for the first time. Today, the Marshes provide less than 20 percent of the fish consumed in the country. Some of this is due to the increase in aquaculture upstream in the Euphrates and Tigris Rivers, but mostly it is a result of increased demand from population growth and the decline in the extent of the wetlands. Most telling is that the number of fish species in the Marshes declined from seventy to ten. Tiamat, the goddess of salty water, was once again exerting her authority.

The Marshes demonstrated remarkable resilience to the major shocks imposed in the 1980s and 1990s. When water returned, so did much of the flora and fauna, although the additional pollution loads from agriculture, municipal waste, and the soil affected the types of species that were able to survive in the Marshes. The greatest problem, however, was the deterioration in water quality, which rendered it undrinkable for humans and some animals. The buffalo population in the Marshes went from 80,000 in 1990 to fewer than 5,000 in 2014.[3]

There are no reliable figures available on how many people were, as a result of the Iraqi regime's decision to drain the wetlands, forcibly displaced from the Marshes or from the towns that depended on them. Estimates range from 300,000 to as many as 750,000.[4] Many former residents established their families in urban centers such as Basra and Nasiriya in the 1990s and were reluctant to leave their new homes and return to the Marshes. In the cities they had access to better education and health services, improved housing, electricity, and safe drinking water. Moreover, their children had never seen the Marshes, let alone lived there. Jassim and his family still had property in Chibayish, and Jassim and his father would visit whenever possible, but their family remained in Babylon. Tahseen's family from Chibayish and Adel al-Maajidy's from al-Mawajid, along with thousands of others, fled Iraq in the early 1990s and eventually resettled in other countries. Others left after the fall of the regime in 2003.

Some people did return, like Najmah or Saddam Shayal, but very few wanted to live in a reed house in the Marshes; they desired a greater measure of social, health, and educational services. There was also an initial lack of economic opportunities. Until the Marshes were firmly reestablished, those activities that fostered a sustainable economy, such as fishing and buffalo herding, would not thrive. Moreover, the pattern of settlements in and around the Marshes began to take a different shape because of changes in the ecosystem. Families now lived on the edge of the Marsh, in towns like Chibayish, rather than inside the Marshes.

Technology also played a role. *Shakhtura*, a small Japanese-built boat that looked like a long mash-huf with an engine, became ubiquitous, and replaced the mash-huf as the main means of transportation in the Marshes. A shakhtura was not only much faster than the mash-huf but could carry almost twice as much weight in people, fish, or reeds. It could even carry water for buffalo, should the need arise. The motor allowed easier access to remote fishing areas and to markets for both the purchase and sale of goods. At the same time, however, propellers stirred up the sediment and tore up small reeds and grasses, and the resulting increased turbidity of the water choked smaller organisms. The peace and quiet of the Marshes were now disturbed by the noise from outboard motors.

Electrocuting fish became common as a means of increasing the size of catch. A small basket that carries an electric charge from a battery in the shakhtura is placed at the end of a long pole. When a current runs from the battery to the basket, all the fish within one meter are stunned or killed, along with most other living beings in the area. Electro-fishing is now illegal in the Marshes, but the ability to catch many fish in a short period of time and the lack of enforcement has allowed the practice to remain quite common.

The Marsh system itself had changed. The extent of the wetlands was substantially less than before the draining. After Saddam implemented his plan to drain the Marshes, only 10 percent of the Marshes remained, which was the section in the northeast of Hawizeh. When the water returned after 2003, there were four distinct, but smaller, marshes: Hawizeh, a much-reduced Central Marsh, and two distinct sections of Hammar Marsh, east and west. A few smaller marshes, such as Abu Zareg, returned as well.

Agriculture and oil development occupied a significant portion of the former wetlands. South of the Euphrates River, the wetlands east and west of Chibayish had been converted to farmland, as had a northern section of the Central Marsh. Saddam had given this land to wealthy landowners who supported him, and who were powerful enough to impede marsh restoration and keep the land for agriculture. Oil development dominated parts of eastern Hammar Marsh and southern Hawizeh Marsh.

Those areas of marsh that were reflooded after 2003 evidenced a remarkable ability to return to productive ecosystems. Nevertheless, troubling signs remained. The initial euphoria surrounding the reflooding of the Marshes soon gave way to a somber reassessment of the long-term impact of the draining. The narrative switched from the idea of restoring the Marshes, to a focus on saving the Marshes. What was needed was nothing less than a major international effort to preserve the remaining wetlands.

A continued supply of fresh water into the Marshes was crucial to the development of Mesopotamia. Fields of wheat, barley, rice, vegetables, fruit, and date palms lie adjacent to the Marshes and draw irrigation water from the rivers and canals that are part of the Marsh system. Extensive reed pastures are an essential part of this system as well, since they are used in construction and mat weaving, and as fodder for buffalo. The Ma'dan are buffalo breeders first and foremost, and their livelihoods depend on young reeds and fresh water. Buffalo not only need the water to drink, but also to thermoregulate their bodies in the summer heat. Therefore, when the water becomes salty and polluted, it not only affects humans, but the entire ecosystem. Buffalo can tolerate water up to 6,000 parts per million of salt content, but no higher.

When the Ba'th regime drained the Marshes in the 1990s, the buffalo breeders were forced to migrate to central Iraq, locating in cities such as Hilla, Samarra, Fallujah, and Balad that had water channels and drains. In the process, they lost 90 percent of their herds, with buffalo dying or being sold off by their owners to raise money for the uncertain journey ahead.

Sayed Ismail was one of the Ma'dan who were displaced from the Central Marsh. He was far from alone—not only did the human population in and around the Marshes decline by more than 80 percent, but the buffalo were decimated as well. Ismail moved out of the Central Marsh with only two buffalo, out of an original herd of sixty, to Balad, 500 kilometers to the north.

After the Central Marsh was reflooded in 2003 and 2004, Ismail and many of his friends moved back to the Central Marsh, hoping to return to their traditional way of life. But the Marsh had changed. There was water, but it was salty. And then there wasn't water—or very little, due to extreme drought conditions. In 2009, 2015, and 2018, the water level of the Euphrates River at Chibayish, where channels and streams feed the Central and Hammar Marshes, dropped from over one meter to less than three-tenths of a meter. What water was left in the Marshes was saline, up to 12,000 parts per million, which is too salty for humans to drink. It is also too salty for the buffalo. Thus, the animals would leave their sheds in the morning to forage for young reeds in the Marshes but then could not drink the water. The closest fresh water that was suitable for the buffalo was in the Euphrates River, twelve kilometers away.

Sayed Ismail lived in a traditional reed house in the middle of the Central Marsh, built on the site of an ancient village called Eshan Kabbah. He built his herd back up to sixty buffalo, but the drought years were difficult for him. He managed, however, with the help of a shakhtura. In the middle of his boat sat two large plastic tanks that can hold 1,000 liters of water each. Next to them was a small gasoline-powered pump that could fill the tanks with water from the Euphrates River, which was less saline than the water in the marsh. Ismail drove his boat twelve kilometers from his house to the river, filled up the tanks, and then returned. He then transferred the water to a brick-and-concrete tank in front of his home. The young buffalo drank first, and then the older ones. He repeated this six times a day. During wet years, the water is fresh enough for the buffalo to drink. During drought years, however, the amount of water Ismail brings from the Euphrates is barely enough for the buffalo to survive. And although Ismail doesn't think about the future, he knows the climate is changing. The salt tells him.

On October, 10 2002, Salam, Tahseen's younger brother, was released from jail and given amnesty after serving ten years of his twenty-year term. He joined thousands of other political prisoners who were set free. It was an attempt by Saddam to appease those in the international community, who condemned Iraq's record on human rights (the government's recent policy of cutting off the tongue of anyone who publicly criticized the regime did not sit well with human rights advocates), at a pivotal moment when the U.S. was threatening war. Saddam also thought releasing prisoners would bolster his support at home in the face of mounting criticism—and not just political prisoners at that. Almost all prisoners, except spies, were freed. It was a bold move. Even those convicted of murder were let go on the condition that they seek forgiveness from families of their victims. The criminals were back on the streets.

Eighteen months later, Saddam had been deposed and the U.S. and its coalition partners were officially in charge, although not necessarily in control. The U.S.-led Coalition Provisional Authority (CPA) assumed control over all branches of government to ensure a smooth transition to a democratically elected parliament. Thousands of troops deserted the Iraqi army during the invasion, and the new American administrator of the CPA, Paul Bremer, soon disbanded the rest of the force. 300,000 Iraqi soldiers left with their guns and whatever they could loot from army bases and returned to their homes. They no longer had a salary or a pension and faced an economy that was almost destitute. It was a recipe for disaster.

Infrastructure throughout the country was in abysmal shape after the bombing, with most water treatment plants destroyed. Iraqi dinars had very little value. Food and bottled water, on the other hand, were almost priceless. With the police also in disarray, crime and corruption became the norm, and the country quickly descended into chaos. The invasion may have toppled Saddam and the Ba'th regime, but it unleashed regional, tribal, and religious factions that had been suppressed for nearly three decades. Anger at the economic situation and the occupying forces first turned to protest and then violence. The government remained disorganized, with some ministries operating and others closed.

The Ministry of Irrigation, soon to be renamed the Ministry of Water Resources, was still operating, but at a reduced level. Dr. Abdul Latif Rashid, a Kurdish engineer and member of an opposition group in exile to the Iraqi government, was appointed minister. Jassim returned to work at the new ministry two weeks after the occupation of Baghdad, as did most of the employees in the Ministry of Water Resources. With multiple ministries sharing space in the Ministry of Oil building, staff were forced to alternate shifts due to space limitations.

The Marshes, however, were not at the top of the national agenda in the days following the invasion. The most pressing issue for the CPA was how to reduce or eliminate the influence of the Ba'th Party in the new government. The U.S. decided to fire all government employees who had been Ba'th Party members and ban them from future employment in the public sector. Critics decried this policy, since many government professionals, including professors, doctors, lawyers, and engineers, had been members of the Party in name only; they played no role in decision-making and were often forced to join the party to keep their jobs. Over thirty thousand senior professionals were fired under the de-Ba'thification policy. This likely would have been Jassim's fate as well, had he acceded to the threats and torture twenty-five years earlier and joined the Party.

There were many levels of Ba'th Party membership. The de-Ba'thification policy was soon amended to allow those who had no decision-making role to retain their positions. However, many decided not to return and either stayed at home or pursued employment in the private sector. Supervisors and party leaders remained barred from working in the public sector. Three years later, the policy was again changed to allow some members of this senior group to return. Few responded since by this time they were working elsewhere. The loss of key professionals made for a rocky transition for the new government. The de-Ba'thification policy also ensured that there was a large, educated, and well-equipped group of people that could work against the new administration—and they did.

The Ministry of Water Resources (MoWR) was one of the few government ministries able to continue working through the chaos of 2003 and 2004. It had been severely underfunded in the last few years of Saddam's rule, with an annual budget of only one million U.S. dollars in 2003. In 2004, this was increased to $150 million. The MoWR was fortunate, as many ministries had no place to work and no resources at their disposal.

The CPA disbanded in June 2004 and transferred authority to an interim government led by Prime Minister Iyad 'Allawi. The goal was to hold national elections in early 2005. Nevertheless, the transfer of power did little to stem either the violence or anti-American protests in the country. Because the de-Ba'thification policy had decimated the Iraqi army, coalition countries had little choice but to keep their troops in Iraq to stem the tide of insurgent groups. This was met with violent resistance from a variety of armed factions, including Kurds in the north, a Sunni army in the central part of the country, and a Shi'a army in the south. Groups that had been loyal to Saddam, such as the Fedayeen Saddam, were part of the insurgency forces fighting the government. There was the Mahdi Army that was loyal to the Shi'a leader Muqtada al-Sadr, and Jama'at Ansar al-Sunna—a largely Sunni group that included Kurdish fighters—among others. They fought against the government. They attacked the U.S. and coalition forces. They attacked each other.

In August 2003, a terrorist group with links to al-Qa'ida set off a bomb at the United Nations headquarters inside the Canal Hotel in Baghdad, killing 22 people and injuring 100. One of those killed was the UN High Commissioner for Human Rights and Special Representative to Iraq, Sergio de Mello. The bombing sent a chill through the UN system. Hassan Partow, author of the original UNEP report on the Marshes, had met with de Mello only three days before. Hassan was working out of a temporary office in the hotel on a post-conflict environmental assessment of Iraq for the UNEP. On the day of the bombing, he left his office for a meeting with one of the appointed CPA governors at the Iraqi parliament building. Ten minutes after exiting the Canal Hotel, he heard a loud explosion and turned to see a cloud of black smoke rising from the UN offices. Three days later, when he was allowed back into his office to collect some papers, he saw de Mello's devastated office.[5] A second bomb went off at the Canal Hotel a month later.

The elections in January 2005 saw a shift in political power at the national level. An Islamist coalition called the United Iraqi Alliance (UIA), comprised of the Supreme Council for Islamic Revolution in Iraq (SCIRI) and al-Da'wa, garnered 50 percent of the vote. The new prime minister was Ibrahim al-Ja'fari from al-Da'wa. Twenty-five years before, Tahseen, along with countless others, had been executed merely for being a member of al-Da'wa.

Iraqi politics might have been changing, but the new political realities did little to stem the tide of violence in the country. Public services had been curtailed, inflation continued, and conflicts among Kurdish, Sunni and Shi'a factions intensified. A new constitution was approved by the electorate in 2006, despite dissent from the Sunni parties. The constitution explicitly stated that "no law which contradicts the undisputed rules of Islam may be established and no law which contradicts the principles of democracy may be established."[6] Whether these two statements were in conflict remained to be seen. What was apparent was that the clergy could now pass judgement on new legislation. Iraq now had a constitution and a democratically elected parliament. Nuri al-Maliki, also a member of al-Da'wa, became prime minister. Meanwhile, the country nearly descended into civil war.

One positive outcome of the constitution and the new elections was that women were legislated to hold at least 25 percent of the seats in the

National Assembly. Socially conservative Islamists, however, had different ideas. They passed a resolution to abolish a 1959 law on the personal status of women, and wanted all family matters to be decided by sharia (religious) courts. The resolution was repealed two months later in response to protests by women's groups, but the social conservatives had made their point. In the midst of increased violence in the country, prominent women and women's organizations became targets of insurgent groups. The enforcement of dress codes, which were relaxed under the Ba'th Party, became more common.

In the latter half of the twentieth century, women in Iraq assumed prominent roles in law, medicine, teaching, and academics. There were minimal restrictions on their activities and on how they dressed. This changed after 2003, particularly in the south. Women who once dressed as many professionals might in Canada, Britain, or the U.S., now were wearing a hijab. As the former Dean of the Law School at University of Basra noted, "Saddam is gone now for more than three years, and we are glad to be rid of him. But if you came to my office five years ago, you would see me in western dress like in Canada, with no scarf on my head. Now you see me in a hijab. I think if you come back to see me five years from now, I will be wearing a niqab. This is not progress and not democracy."[7]

Figure 23. Traditional fishing in Umm al-Niʻaj (courtesy of Mootaz Sami).

The escalating violence in the country continued into 2006 and 2007 and caused a massive dislocation of people, particularly in the central and southern parts of the country. Over two million people, out of a total population of thirty million, were internally displaced between 2003 and 2007, and an equal number fled the country entirely.[8] Two to three thousand refugees a day poured over the border into Syria in 2006 alone.

Jassim was driving between his work in Baghdad and his home in Hilla each week, often along the main road between Baghdad and Basra. There were U.S. troops posted all along the road, and he had to endure many checkpoints. Iraqi cars and trucks were never allowed to pass U.S. Army vehicles on the road, and it was common for a stretch of highway to be closed for two or three hours because improvised explosive devices, or IEDs, had exploded in the area. Travel was the most dangerous in 2006 and 2007. Sometimes, when Jassim drove with his family back to Chibayish, the bridge over the Euphrates River north of Nasiriya would be closed due to security concerns. He had to convince whoever was guarding the bridge to allow them across, or else they were forced to return the 200 kilometers back to their home in Hilla. Most of the time there were Eastern Europeans guarding the bridge and they would allow Jassim to pass if he went quickly.

6

A CAUTIONARY TALE OF RESILIENCY

The reflooding of the Marshes continued, despite the context of violence and a pending humanitarian disaster due to population displacement. It was a hopeful sign in what was otherwise a bleak period in the country's history. When Iraq was invaded in 2003, the area of marshland inside the country was less than 10 percent of its extent in 1973, or roughly 1,000 km². By 2006, largely due to local initiatives, the size of the Marshes had increased to nearly 4,500 km² (figure 24). The plight of the Marshes and the Maʿdan was now international news, and donor countries were quick to offer help. UNEP, Italy, Canada, and the U.S. provided millions of dollars to help with hydrological modelling, training in wetlands biology, water resource management, and governance. More important, however, was the willingness of the Iraqi government to put the restoration of the Marshes higher on the political agenda. Part of this process was to create a new Ministry of Environment in 2003. From 2004 to 2010, the minister was Narmin Othman—a former member of the Peshmerga, the military forces of Kurdistan who were outlawed under Saddam's regime. She proved to be a tireless advocate for the environment and for the Marshes. A second important step was to have the Council of Ministers approve the Center for the Restoration of the Iraqi Marshes (CRIM—soon to be CRIMW, when "wetlands" was added to the name). This happened in February 2004. It sent a signal to Iraq and the world that marsh restoration would be a focus for the new government.

The reality of any government is that individual departments and ministries are often in conflict over power and human and financial resources. With 95 percent of the country's export earnings coming from oil, the

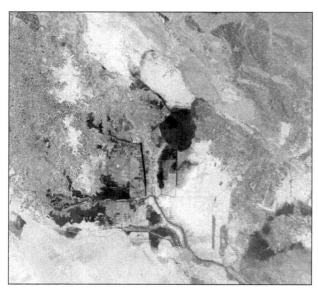

Figure 24. Satellite image of Marsh extent, 2006. (MODIS images courtesy of the U.S. National Aeronautics and Space Administration).

Ministry of Oil was by far the most powerful ministry in Iraq, outside of the military and security apparatus. The Ministry of Environment, on the other hand, had very little influence and was subservient to the ministries of oil, agriculture, and water resources.

CRIM experienced similar problems. At its inception, the center was a small office of only four staff within the Ministry of Water Resources. However, it wouldn't remain that way for long. The first director, Hassan al-Janabi, was well respected, knowledgeable about what happened to the wetlands, and committed to seeing the Marshes recover. He also had strong political connections. More importantly, he and a few others recognized that a strong voice from civil society was needed to keep environmental issues and the Marshes in the public consciousness, and to ensure that the government delivered on its promises. The problem was that there had been no environmental non-governmental organizations (ENGOs) in the country during the Ba'th regime, and no organized voice for the Marshes.

The 2001 UNEP report on the Marshes shocked not only environmentalists but the expatriate Iraqi community as well. One of those particularly affected was Azzam Alwash, an Iraqi-American engineer living in California who had grown up in Nasiriya and spent time in the Marshes during his youth. Azzam was on the board of directors of the Iraqi Foundation, a group that advocated for human rights and democracy in Iraq and had strong ties to the U.S. government. Over the next few years, Azzam worked to bring international attention to the plight of the Marshes. When the Iraqi regime was toppled in 2003, he moved operations to Iraq and started Nature Iraq, the first ENGO in the country, with assistance from both the U.S. and Italian governments. Nature Iraq filled a crucial role as a link between the government and the people living in, and near, the Marshes, as well as in obtaining international funding and support for marsh-related initiatives.

Iraq had no shortage of excellent scientists and engineers, both inside and outside the country. What was missing in 2003 were organizations that could work together and be advocates for the Marshes and the people, both internationally and within the government, and that could also deliver on projects to restore—or at least help rehabilitate—the Marshes. The Center for the Restoration of the Iraqi Marshes, the new Ministry of Environment, and Nature Iraq provided this framework. There was also an irrigation engineer who believed in the cause—one who was knowledgeable about the daunting engineering challenges that faced the Marshes and, more importantly, who understood the people and communities in the region . . . Someone who would be able to provide a link among the government, Nature Iraq, and the local population. That person was Jassim al-Asadi.

As part of his work at CRIMW, Jassim spent much of his time visiting local communities and talking to people throughout the Marshes about the changes in the wetlands. He understood the difficulties that Marsh dwellers had experienced over the previous two decades. His family had been displaced and suffered as a result, and he was determined to improve the lives of those living in the Marshes. Jassim acted as a crucial link between the engineering solutions and the people.

Despite the unsettled situation in Iraq in the years following 2003, billions of dollars were provided from international donor institutions for Iraqi reconstruction. Who actually benefited from these funds is very much in question, but what is clear is that the general population continued to suffer. Iraq might have been considered a country on its way to peace and democracy, but activities on the ground told a different story. Between 2003 and 2008, ninety-four aid workers were killed, 300 were wounded, and eighty-nine were kidnapped. Much of the funding for reconstruction went to foreign contractors, healthcare continued to regress, and very little of the money went to alleviate suffering in the poorest segments of society.

Restoring the Marshes, while less of an immediate concern than rebuilding water and sewage treatment facilities or transportation systems, was high on the agenda of a few countries, in particular Japan, Italy, and Canada. The problem was that within Iraq, every ministry needed funding—and each ministry had a different mandate. For instance, CRIMW was interested in the physical process of reflooding the Marshes; the Ministry of Environment, with a national mandate, was focused on environmental monitoring, pollution control, biodiversity, and ecosystem preservation; the Ministry of Oil cared about oil development, regardless of the impact on the Marshes; the Ministry of Municipalities and Public Works wanted to ensure that people had jobs and incomes, and that water and sewer facilities were restored. Indeed, there had been no planning for what a post-Saddam Iraq would look like, and, for the moment, chaos reigned.

It would be years before there would be cooperation between the Ministry of Environment and the Ministry of Water Resources over the Marshes. In the meantime, for the Marshes at least, chaos was not a bad thing. Water had started to return to many areas, in large part due to local citizens' initiatives. The framework for a comprehensive restoration plan for the Marshes, developed by the USAID in 2004, involved cooperation among multiple ministries. Jassim and the staff at CRIMW then worked on implementing the restoration plan with Nature Iraq. The initial component was a series of water-control structures to regulate the water levels in the Marshes. The days of natural flooding—the pulse of water that cleansed the ecosystem—were over; water management was now crucial.

The plan had three components. First, survival of the Marshes depended on ensuring a controlled flow of water to key regions of the wetlands. CRIMW therefore installed what are known as "feed regulators" to allow water from the Tigris River to flow into the Central Marsh; these were completed in 2012. Regulators were also built on the Euphrates River. The main purposes of the feed regulators were to ensure reliable irrigation water for the region's farmers, and to provide needed water to the Marshes.

The second component involved building an embankment on the Euphrates River, east of Chibayish, to keep river levels high enough to both reflood the Marshes and maintain an acceptable level of water quality. The embankment acted like a weir, backing up the flow of water until levels were high enough to overflow the embankment and flood the Marshes.

Figure 25. Young girl poling a *mash-huf* (courtesy of Jassim Al-Asadi).

This was completed in 2010, despite concerns from the Basra Governorate that the embankment would reduce water flow and water quality downstream of the Shatt al-Arab River. Their concerns were unfounded, and increased flow along the Tigris more than made up for the difference.

The third project was one proposed by Azzam, Jassim, and Nature Iraq; it involved diverting water from the Main Outfall Drain (M.O.D.) into West Hammar Marsh. The M.O.D. had proven to be effective in removing salts and other pollutants from upstream agricultural areas and transporting them to the Gulf. In the early 1990s, the Iraqi government adapted the M.O.D. to help drain the Marshes. It was a radical suggestion to now use water polluted by agriculture runoff to help replenish West Hammar Marsh. Nature Iraq completed an environmental impact assessment on the proposed project for the Council of Ministers, and the proposal stipulated: establishing a feeding channel that contained a regulator at the front of the inlet to control the incoming water; establishing a phytoremediation treatment system before using water from the M.O.D.; and ensuring that Euphrates River water is mixed with the water from the M.O.D. to reduce the salinity of the water entering the Marshes. Unfortunately, however, the Ministry of Water Resources did not implement all the recommendations due to the perceived negative health impacts.

Wetlands have been used to provide tertiary treatment of pollution in a few countries, but dumping virtually untreated waste into a marsh was something new. Still, there was simply no choice: the Euphrates River did not provide enough water to revive the Hammar Marsh. Without the M.O.D. water, the Marsh would become a desert or turn into an anoxic, briny wasteland. Thankfully, the project was a success. The water might not have been fit for human consumption, but birds, fish, and reeds returned. Buffalo production began anew, with positive impacts for the local economy. Although the quality of the marsh needed to be monitored continuously, the experiment worked. As Jassim liked to say, "Many people outside the Marshes prefer the taste of the fish caught in water from the M.O.D. They come pre-seasoned!"

Jassim was slowly becoming a leading advocate for the preservation of the Marshes, both nationally and internationally. As part of his work for the CRIM, he carried out engineering and environmental assessment projects and worked closely with local communities in and around the Marshes. He gave lectures on the Marshes at national and international conferences, conducted archeological surveys of the area, organized workshops, and promoted environment and development projects. Jassim also increased his cooperation with Nature Iraq, edited a column titled "A Window on the Environment" in a Baghdad newspaper, and supported the development of veterinary centers for Marsh dwellers.

Restoring the Marshes, however, required more than additional water and engineering structures; it needed marketing. This involved convincing both the Iraqi people and government agencies that the Marshes were a unique resource worth protecting. Engaging the international community was part of this, both in terms of receiving necessary financial assistance and recognizing the importance of the Marshes and the people living there. Most of all, an effort was needed to help people living in the Marshes understand the changes taking place, and to support them by advocating for improved health and education services; Nature Iraq was crucial for both.

Two decades of war, international economic sanctions, and severe restrictions on Iraqi citizens effectively isolated Iraq from the international community. Although Iraq was one of the founding members of the United Nations, its participation in international conventions was almost non-existent. This started to change in 2004 when Iraq began negotiations to join the World Trade Organization. Indeed, one of the least contentious mechanisms available to engage, or reengage, in the international community is by becoming a member of various UN environmental conventions, such as the Convention on Biodiversity and the Ramsar Convention on Wetlands of International Importance.[1] The restoration of the Marshes provided a perfect opportunity for Iraq to participate in these conventions.

The UNEP was already committed to the Marshes, both having published Hassan Partow's 2001 report and undertaken a post-conflict environmental assessment of the country in 2004. Nature Iraq began increasing awareness of the Marshes through the UN sustainable development conferences in 2004 and 2005. The Ministry of Environment, CRIM, and Nature Iraq soon set their sights on having Hawizeh Marsh designated as a wetland of international importance under the Ramsar Convention. There are over 2,300 Ramsar sites around the world, and by 2007 Iran had twenty-one designated sites, while Iraq had none.

To become a member of the Ramsar Convention, a country must designate at least one wetland as a potential site. This site must meet one or more of nine criteria set forth in the Convention. Once it becomes a member of Ramsar, the country is then obliged to manage the wetland to ensure that ecological communities are maintained. The issue of gaining Ramsar designation was important enough that CRIM changed their name to CRIMW, the Center for the Restoration of Iraqi Marshes and Wetlands, in 2007. The northern section of Hawizeh Marsh, one of the few wetlands that withstood the devastation of the 1990s unscathed, met five of the nine Ramsar criteria and was placed on the List of Wetlands of International Importance in 2007. Iran did not request the same Ramsar designation for their part of Hawizeh, known as Hor al-Azim, because they had already begun draining this section of marsh for oil production. People had been resettled, and a series of dams and dikes reduced the flow of water to key oil production areas.

Three other Iraqi wetlands have since been designated Ramsar, including the Central Marsh and the East and West Hammar Marshes. In 2010, the Iraqi government added the Hawizeh Marsh to the Montreux Record (part of Ramsar), which acknowledges that changes in the wetlands have occurred—or are likely to occur—as the result of development, pollution, or other human interference. The Montreux Record is not a list countries aspire to be on; being included sends a strong message that after having survived the devastation of the 1990s, Hawizeh was again under threat.

Becoming a member of the Ramsar Convention and designating Hawizeh as a wetland of international importance were significant in getting the Marshes recognized around the world. But the crucial need was to convince the Iraqi people and, more importantly, the Iraqi government that the Marshes should be preserved, even with the ongoing pressure for more agriculture and oil development. Designating at least part of the Marshes as a national park, which would require a decision by the powerful Council of Ministers, might have this effect.

Nature Iraq and the Italian Ministry of Environment, Land, and Sea believed that a national park was an essential step toward preserving the Marshes. They worked with the Iraqi ministries of Environment, Water Resources, and Municipalities and Public Works to develop a proposal for a Mesopotamian Marshlands National Park (MMNP). The process of developing a plan and acquiring government approval took eight years, but was ultimately successful.

The MMNP was intended to help restore and preserve the Marshes and its cultural heritage, and to ensure the sustainable development of the surrounding area. Developing a plan for the park required technical work on water quality, hydrology, and biodiversity as well as workshops to inform community and district leaders and the public of national park benefits. Jassim played a unique role in this process. His life had now come full circle, from a childhood of living in and being part of the Marshes, witnessing its destruction at the hands of Saddam Hussein, and now being at the forefront of helping protect the ecosystem. Because of his roots in the Marshes and his technical expertise as an irrigation engineer, other Marsh dwellers trusted him. He also understood that a cooperative approach was needed to ensure the sustainability of the remaining wetlands. Jassim played a pivotal role organizing and facilitating many of the community workshops that were needed to inform people about the national park initiative. He was back in the Marshes and making a difference.

The initial strategic plan for the national park was presented to the Iraqi Council of Ministers in 2010. Unfortunately, the key ministries of oil and agriculture were not interested and refused to approve the proposal. It didn't help that the Ministry of Environment and the Ministry of

Water Resources were at odds as to who should be the lead agency on the Marshes. Their mandates were different, and the Ministry of Environment was still in its nascent phase, unable to exert pressure on the more influential economic development ministries to convince them that a national park would benefit the entire country.

Two years later, however, there was a different Council of Ministers. This time, the key ministers were convinced that a national park would have minimal impact on agriculture and oil production and might stimulate tourism. In 2012, the council approved the proposal to establish the Mesopotamian Marshlands National Park, with a focus on the Central Marsh because of its accessibility, ecology, and cultural diversity. Nevertheless, lack of a coordinated effort by the Ministry of Environment and the Ministry of Water Resources, along with some reluctance by the three governorates involved (Dhi Qar, Maysan, and Basra), hindered the implementation of the park's management plan. Today Jassim is a member of the committee overseeing the park, but has thus far been displeased with the progress. While there is a national park on paper, few people in Iraq are aware of it, and little has been done to develop the park's infrastructure.

It was a warm summer evening in Istanbul. Jassim was relaxing and reading in his sixth-floor hotel room in the heart of the city, when he suddenly heard the *chop-chop-chop* of helicopter rotors and the sound of gunfire in the distance. He rose from his chair and opened the window curtain. Military helicopters were flying overhead, and a large crowd of people had gathered on the streets below. As he gazed at the scene unfolding before him, his cell phone rang. The excited voice on the other end was that of another member of the Iraqi delegation to the UN meeting telling him that there was an attempted coup taking place, and that Turkey's President Erdoğan had ordered the army into the streets. Just after he hung up, a text message arrived from the organizers of the meeting informing him that the discussions the next day would be delayed and possibly canceled. It was July 15, 2016, and the UN Education, Scientific, and Cultural Organization (UNESCO) was scheduled to meet and take a vote the following day on whether the Marshes would become a World Heritage site. It was a crucial step toward preserving the Marshes that had been years in the making—the last thing Jassim and the Iraqi delegation wanted was a delay in the proceedings.

As he watched the scene play out in the streets, there was a knock on the door. One of the hotel staff members requested that Jassim follow him downstairs to a safe room; hotel management was unsure of what was transpiring outside and wanted all guests safe and accounted for. Jassim joined other guests who emerged from their rooms and made their way to the basement safe room, but then he had second thoughts. He turned to the staff member in charge, informing him that he needed to return to his room to collect a few things and would be back shortly. Jassim had no intention of returning to his room. He exited the building and joined the people in the streets to find out what was happening. Istanbul was nothing like Baghdad had been thirteen years before when Saddam Hussein was toppled, and so he was unafraid. As the crowd made its way to the bridge over the Bosporus Strait that connects Europe to Asia, it was cut off by government security forces dressed in riot gear, with guns drawn. A small faction of the military was attempting to take control of key government sites in Ankara and Istanbul, including the entrance to the bridge over the Bosporus. What began as a peaceful, if noisy, protest in support of the coup was now turning deadly. After seeing the armed security forces, Jassim decided it was better to learn about the proceedings on the news rather than be an active participant. The UNESCO issue was more important— or at least it was to Jassim. This turned out to be a wise decision: during the coup, over 300 people were killed and 2,100 injured.

The UNESCO meeting pertained to the UN Convention Concerning the Protection of the World Cultural and Natural Heritage, which entered into force in late 1975. As of 2023, 191 UN member states have signed the convention. The aim is to preserve unique natural and historical sites around the world for the public benefit. Areas that meet a set of cultural and natural criteria can be designated as World Heritage sites by UNESCO; these sites are then accorded legal protection under the convention and are also eligible for funding from the World Heritage Fund.

There are over 1,100 World Heritage sites around the world. Applying for and receiving World Heritage designation is both a time-consuming and costly procedure, leading to some criticism that the program is biased against poorer countries. But there is little question that having the designation increases both international exposure and tourism revenues. As a result, there is much competition for designated status. The idea of registering the Marshes on the list of World Heritage sites arose in 2004, when the Iraqi

Ministry of Culture, Tourism, and Antiquities submitted a preliminary list to UNESCO of possible sites in Iraq. The Marshes was one of them.

In 2007, staff from UNEP and UNESCO met with the Ministry of Environment staff in Baghdad to draft a proposal for an international cooperation project focused on getting World Heritage designation for the Marshes. The Iraqi government approved the draft in 2009, but the project was soon mired in controversy. Ministries had competing interests, and key people involved with the Marshes—including Hassan al-Janabi and Jassim—were left out of the high-level discussions. To make matters worse, Iran and Turkey were firmly against the proposal.

Three major problems arose immediately. First, the IUCN (International Union for Conservation of Nature) and ICOMOS (International Council on Monuments and Sites)—both official advisors to the UNESCO World Heritage Committee—felt that the Marshes did not adequately meet the World Heritage guidelines. They submitted a report to the Iraqi Committee in May 2016 that criticized the application file. Second, neither Turkey nor Iran had any interest in having the Marshes designated as a World Heritage site, believing it would force them to provide extra water to sustain the Marshes. Lastly, and worst of all, the committee managing the Marshes file for Iraq consisted of high-level bureaucrats who knew little, and cared less, about the Marshes.

The governorate of Dhi Qar, in the meantime, formed its own committee to promote the Marshes proposal to UNESCO, with Jassim as a member. The chair of the Dhi Qar committee was Governor Yahia al-Nasseri. Jassim felt that the national committee managing the Marshes file did not have the expertise to ensure a successful vote by the UNESCO committee in the upcoming Istanbul meeting. Al-Nasseri agreed and asked Jassim to compile a list of key people who should be on the national committee. At the top of Jassim's list was Hassan al-Janabi, who was at the time Iraq's ambassador to the UN Food and Agricultural Organization.

Al-Nasseri presented the list to the prime minister and pleaded with him to change the composition of the national committee if he wanted the UNESCO proposal to succeed. The prime minister acceded, and al-Janabi was named chair. Jassim didn't feel comfortable putting his own name forward to be a member, but it didn't matter. With less than six weeks until the Istanbul meeting, al-Janabi's first decision was to add Jassim as a committee member.

Iraq's proposal to UNESCO was termed a "mixed site" approach. Seven properties or locations were incorporated in the proposal, including four natural sites and three cultural/archeological sites. The four natural sites were sections of the Hawizeh, Central, East Hammar, and West Hammar Marshes, all surrounded by buffer zones to ensure the sites would not be impacted by development. The three cultural sites included the former cities of Ur, the birthplace of Abraham; Uruk, originally on the Euphrates River and the largest city in the world during the fourth millennium BCE; and Eridu, the first city in southern Mesopotamia and the home of Enki, the god of fresh water.

ICOMOS and IUCN worried about whether the sites' boundaries ensured their integrity, protection, and management. The first task of the new Iraqi committee was to address these issues, and to dialogue with the two groups to convince them that their concerns were unwarranted; Jassim and al-Janabi were crucial in this effort. Representatives from the two organizations met with them and, after a lengthy discussion, dropped their objections to the Iraqi file. The Iranians, on the other hand, had different concerns. They did not want Hawizeh Marsh included in the file because it was a shared marsh between Iran and Iraq. Additionally, they did not want the name Hawizeh used, since it was an Iranian name. A third issue was whether the convention would obligate Iran to provide water to Iraq. Lastly, they objected to Iraq's using the term Arabian Gulf (which had appeared twenty-six times in the original application) rather than Persian Gulf.

The Iraqi prime minister invited a high-level delegation of Iranians to visit Baghdad and meet with al-Janabi and his negotiating team—which included Jassim—to discuss their concerns. The Iranian group included a representative from the office of Ali Khamenei, the supreme leader of Iran. He would move in and out of the meeting, calling Khamenei's office for advice as the negotiations continued. The discussions went on for twelve hours before a resolution was reached. Iran dropped its first two objections, and all agreed to use the neutral term "the Gulf," rather than designate it Arabian or Persian. During the discussion of the Marshes file at the UNESCO meeting in Istanbul, the Iranian ambassador to UNESCO enthusiastically supported Iraq's application and even read a poem reflecting the positive spirit of their negotiations.

Turkey posed a more serious concern. The government was convinced that designating the Marshes a World Heritage site would obligate them to

Figure 26. Celebrating the UNESCO decision (Jassim with arms raised high) (courtesy of Jassim Al-Asadi).

provide more water to the region. The Turkish ambassador to UNESCO threatened to boycott the session in which Iraq's file was to be discussed. Attempts by the Iraqi delegation to change Turkey's mind proved to be fruitless and discussions stalled. Turkey appeared ready to raise strong objections to the Iraqi proposal at the Istanbul meeting; if this happened, there was little chance of success for the Marshes file.

However, everything changed with the coup. It was as if Enlil—the god of wind, air, earth, and storms—upset at the impasse, intervened. President Erdoğan was in the process of forcefully suppressing the rebellion and the Turkish government appealed for international endorsement for their actions. One condition of Iraq's support for Turkey was that latter drop its objections to the Marshes proposal. On July 16, 2016, after a short delay in the proceedings, the Marshes file came before the UNESCO World Heritage Committee. The Turkish ambassador to UNESCO raised no objection. With vocal support from Iran and most other nations, the Iraqi proposal passed the UNESCO Committee. The Iraqi Marshes were now a World Heritage site.

The combination of war, government restrictions, cost, and a difficulty obtaining international visas severely limited Jassim's travel outside Iraq for most of his life. Nevertheless, he has visited a few of the great cities of the Western world, including New York, Rome, and Istanbul. His love of art and architecture comes alive when he travels to such places. Jassim's favorite city in the world is one surrounded by water, where boats are the main source of transportation—a city of islands and canals, which traces its roots to people fleeing invading armies almost a thousand years ago. While this sounds much like Venice, Italy—a city that he is very fond of—Jassim's favorite city is Chibayish, the place where his family settled when they left Babylon and where he spent his childhood. Little more than a village when he was born, it then grew to a bustling small city with over 60,000 people before being decimated when the Marshes were drained. And like the Marshes that surround it, Chibayish has shown remarkable resiliency, and the population is now once again above 60,000.

Chibayish extends like a snake along the northern bank of the Euphrates River, forty-five kilometers west of Qurna where the Euphrates meets the Tigris to form the Shatt al-Arab River. It lies in the governorate of Dhi Qar, with its capital of Nasiriya, 100 kilometers to the northeast. North of Chibayish lie the vast Central Marsh and to the south the Hammar Marshes. Like Venice, it is a city surrounded by water. The only way out in the 1950s was by a steamboat known as GM (after its engine, manufactured by General Motors in the U.S.), or by paddling one of the small and ubiquitous mash-huf for hours. The first road that connected Chibayish to Nasiriya was built in 1960. The road in the opposite direction—to Qurna—was built ten years later.

Like Venice, Chibayish was constantly under threat from outside forces. Once the tribes associated with al-Muntafiq began to disrupt commercial traffic on the Tigris River in the mid-1800s, the Ottomans embarked on a plan to convert marshlands to agriculture and force Marsh Arabs into peasant status. Co-opting tribal leaders, setting up government administrative districts, and extracting more tax revenues from the local population were key components of the plan. This led to the establishment of a government district over much of the Marshes in 1857 and the appointment of a sheikh—no longer would it be a hereditary position—for tribal federations such as the Muntafiq. Not all the tribes of the Muntafiq accepted Ottoman rule. The Bani Asad was one that resisted. This led to the assassination of Khayoun al-Khayoun, leader of the Bani Asad, by Nasir Pasha al-Saadoun, senior sheikh of the al-Muntafiq in 1866.[2] Al-Khayoun's brother Mohi then became head of the Bani Asad. Mohi was the first of the al-Khayoun sheikhs to cooperate with the Ottomans, who quickly named him the district mayor. It appeared the reign of the tribes in the Marshes might be ending.

Mohi, in turn, was killed in tribal conflict by another brother, Hassan al-Khayoun, in 1893. It seemed stability in Chibayish, at least for the Bani Asad, was elusive at best. Hassan continued to cooperate with Pasha al-Saadoun (now al-Ashkar), and the two parties agreed to build an embankment along the southern bank of the Euphrates, ostensibly for flood protection. Its main purpose, however, was to drain the Hammar Marsh and convert the land to agriculture, thereby controlling the movement of the population. This would have allowed the Ottomans to levy taxes and keep tribal uprisings to a minimum.

Once Hassan al-Khayoun realized the real purpose behind the embankment, he quickly tired of his relationship with Pasha al-Ashkar. He decided the Ottomans needed to be forced out of the Marshes and initiated a revolution against the Turkish rulers. Thereafter, Hassan was known as the King of the Marshes. The Ottomans retaliated and sent more troops to the region and quashed the rebellion. They eventually established a military garrison in Chibayish and turned it into an administrative center for collecting taxes from the clans, ensuring security of passage and enforcing laws. It also provided a base for expanding Ottoman control over the Marshes. In an effort to ensure peace, Jayed al-Khayoun, Hassan's nephew, was appointed chief of police.

Hassan escaped to Hawizeh Marsh after his defeat at the hands of the Ottomans, but in the process managed to persuade a group of 600 loyal followers to join him. There were still plenty of places to hide in the Marshes. Hassan set up a government in exile and sent his sons and supporters to Chibayish to collect taxes from members of his clan. He then adopted the strategy of the Muntafiq and began harassing the Ottomans by controlling different strategic spots along the Tigris River, extracting taxes from goods shipments and firing on troops moving south. Hassan and his troops then launched a large-scale attack on the Ottoman army moving along the river road between Qurna and Amara. His goal was to pressure the Pasha in Basra to allow him to return to Chibayish. In return, Hassan agreed to be obedient to the government and allow safe passage along the road. He was refused.

The Ottomans retaliated and sent troops to find Hassan, and this time he was forced to flee across the border into Arabistan in Persia. His leadership was then transferred to his son, Salem. Hassan married nine women during his lifetime and fathered nineteen sons and eight daughters. He lived to be over a hundred years old and died in 1912. He was buried in the famous Valley of Peace cemetery in Najaf.

Salem al-Khayoun was born in the Marshes in 1883. He learned reading and writing from a cleric in the al-Haddadin clan of the Bani Asad by the name of Sheikh Muhammad Hassan Faraj Allah, who later became Jassim's grandfather. Like many Bani Asad, Salem was enthralled with poetry. In particular, he enjoyed the Nabati poems of the Bedouin—poetry of the people, not the aristocrats. Salem loved writing Nabati poems and excelled at reciting them as well. He was also a man of languages; he spoke both Turkish and Persian and, during his later exile to India in 1915, learned English and Hindi as well.

Salem had long harbored an ambition to lead the Bani Asad. He was courageous and demonstrated a willingness to engage in dialogue, often mixing with the local people in his father's mudhif. He eventually won his father's affection and support. But Salem was not the only one of Hassan's nineteen sons who desired to be sheikh of the tribe. Because of his activities disrupting trade along the Tigris, Salem was imprisoned by the Ottomans for four months during a crucial time in the ascension of a new sheikh. While his father's health was failing, Salem was stuck in a cell as his brothers struggled for control of the tribe. He was released from jail just in time, however, and assumed the role of sheikh with the blessing of his father in 1904.

Salem's personality and knowledge of the various clans and tribes and their respective laws allowed him to become a popular mediator in disputes between clans. The Ottomans followed his progress and formally recognized him as sheikh of all the Bani Asad. Salem held the leadership of the Bani Asad until the sheikhdom was officially abolished in 1924.

The Iraqi state was born in 1921, and a new administrative structure was imposed on the entire country. When the sheikhdom was abolished, Salem al-Khayoun made one more attempt to reinstate the authority of the tribes. He instituted an unsuccessful rebellion against the Iraqi government and was arrested and thrown in jail in 1925. Salem spent the next year of incarceration in four different districts before being placed under house arrest in Mosul, far from the Marshes. It wasn't until 1945 that he was able to return to Chibayish. Salem al-Khayoun died of prostate cancer in the American University Hospital of Beirut in 1954. Today, the al-Khayoun mudhif remains the largest and most beautiful in Chibayish.

One major difference between Chibayish and Venice is that the former is surrounded by fresh water while the latter is nestled in the salty water of the Adriatic Sea. For drinking water, Venice had to rely on rainwater or water brought by ferry from the mainland. In Chibayish, this was never a problem. The abundance of fresh water also meant that lush vegetation and wild animals abounded. In Chibayish, the natural environment was the attraction; in Venice it was the built environment, including the grandeur of San Marco Square or the Rialto Bridge. While Venice grew to a powerful Italian city-state, Chibayish remained unchanged for over a thousand years. Unlike Venice, it was completely isolated.

A second major difference is that while Venice attracts twenty million tourists per year, Chibayish attracts very few. Its lack of accessibility, coupled with the security situation in Iraq, severely limit international tourism. Despite its recent growth, Chibayish will never attract the numbers of tourists that frequent Venice; few places in the world do. However, with the Marshes now a World Heritage site, the Mesopotamian Marshlands National Park ready for possible development, and the security situation in Iraq growing more stable, there is cause for some optimism.

On the eastern bank of the Abu Subbat River, near the main road connecting Chibayish and Qurna—between reed forests that alternate from being green and full of life when the water flows, and dry, brittle, and dead in drought years—stands the Martyr's Monument, a large dome over thirty meters high. It is the most dominant feature in Chibayish today, dedicated to those who died in the Marshes during decades of fighting in defense of freedom, democracy, and social justice. Under the dome structure stands a fountain, and both are surrounded by a large courtyard with gardens and corridors. There is also a statue representing the heart of the martyr, with reeds that symbolize arteries, along with a symbolic burial site. Photographic exhibitions and poetry festivals are regularly held under the dome or in the courtyard.

The Martyr's Monument may commemorate the heroes of the Marshes, but it appears an incongruous structure of iron and cement amidst nature's bounty. There is no use of natural materials or integration into the surrounding landscape. The project was proposed and funded by the Iraq Ministry of State for Marsh Affairs, a small office which no longer exists, and designed and built by a Turkish company. The monument was completed in 2012 at a cost of seventeen million dollars, a large sum in a region with high unemployment, high rates of poverty, and poor infrastructure.

I once asked Jassim to tell me the purpose of the monument, what it symbolizes to the region and the nation, and its potential impact on the local environment. Here was his response.

> "When I was a young child, I would pass by my uncle Hajj Abdul Khaliq Abbas' food store on the way home from school. He would often give me chocolates and crystals, and, in my imagination, I thought about his home and store as being filled with political exiles, teachers, communists, nationalists, and Islamists, all free from the oppressive regime . . . We grew up to the sound of bullets, the smell of gunpowder, and the language of killing. I realized that the reeds surrounding our village were a place of shelter for large groups of armed men fighting against the government."

Villages in the Marshes such as Abu Subbat, Abu al-Narsi, al-Siqil, and many others were reduced to ruin as government forces scoured the region for deserters, insurrectionists, members of outlawed political parties, and any other individual or group posing a threat to the regime. "The Ma'dan will never forget the cruelties and horrors inflicted on the Marshes, the people, and the environment," Jassim told me.

Jassim admits that the Martyr's Monument has many features that evoke passion in those who visit. But he feels it is in the wrong place. The monument should have been built in Baghdad, in a central location where it can be seen by people of different nationalities, sects, ethnicities, and cultures—that they might better learn about what happened in the Marshes, so their children and grandchildren will know of these martyrs who fell under the shade of reeds and papyrus fighting for freedom. The people living in the Marshes don't need a monument; they have their experiences and their stories.

The monument also stands in stark contrast to the surrounding ecosystem and to the nearby—but unfinished—Iraqi Marshes Museum. Jassim was on the planning committee for the museum when it was first proposed

Figure 27. Conflicted space: the Martyr's Monument (courtesy of Meridel Rubenstein).

in 2007. The museum is intended to adhere to environmental and local standards, with due consideration for the height of reed dwellings, the visibility of marsh spaces and wildlife, and the water on which it is based. It is expected to offer a seamless integration with the landscape, unlike the Martyr's Monument which overwhelms the reed forests and includes a road that cuts off the movement of water. The first stage of the project was completed before the Iraqi economy collapsed and funding was cut. The project remains on hold. Meanwhile, the Martyr's Monument was erected nearby, much to Jassim's dismay.

In Iraq, such projects that are undertaken without input from the community often waste money and threaten the functional and aesthetic aspects of the surrounding environment due to being misplaced. Jassim concluded the article thus:

> When will we realize the importance of comprehensive planning and integrating the structures of modernity with the natural environment? And when will we give the people of the Marshes green villages with employment that is integrated with the beautiful marsh environment and not something totally alien?

In November 2022, the city of Chibayish was named a Global Wetlands City at the UN's Convention on Wetlands—one of only twenty-five cities that have received this designation. The reason is partly that Chibayish sits on a 300-meter strip of land between the Central Marsh to the north and West Hammar Marsh to the south. Other criteria for being awarded the designation include: ensuring development does not degrade the wetlands, that the wetlands play a role in enhancing human well-being, the inclusion of indigenous and local communities in decision-making, and promoting events around World Wetlands Day. The UN designation is intended to raise public awareness of the benefits of wetlands and encourage wetlands protection in municipal planning.

7

DISTURBING AN OLD SUMERIAN GOD

A bright white expanse stretched farther than the eye could see, studded with desiccated tree stumps and hills of cane huts scattered about, wind rustling through them, midst squawks of ducks and storks as they rose from the water then drifted slowly like clouds above this body of water that lay like an old Sumerian god whose solitude had not been disturbed by mankind for millions of years.[1]

Twenty years after he fought for the Iraqi army in the First Battle of the Marshes during the Iran–Iraq War, Khaled Assam was back in the Hawizeh Marsh, this time as a lieutenant colonel in the police force in charge of border control. There were no longer waves of young Iranian troops charging at him, willing to risk their lives for the Islamic Republic; instead, they had been replaced by smugglers moving arms and drugs across the border, destined for elsewhere in the Middle East. There was one constant, however: Hawizeh Marsh was again under assault. This time, it was not the Iraqi government destroying the Marsh to drill for oil or make it easier to bring tanks into battle, but the Iranian government who was responsible. In 2007, Iran built a sixty-four-kilometer-long, six-meter-high earthen dike in Hawizeh Marsh, just inside its border with Iraq. A few concrete guard towers were constructed at strategic points. Anyone coming too close or trying to tear down the embankment could face a hail of bullets for their efforts. One of the guard towers was directly across from a site in Hawizeh where Kahled watched for smugglers. Graffiti painted on the tower read, "Remember Hussein's Thirst."

The border between Iraq and Iran cuts through the northeastern section of Hawizeh Marsh and was part of the area Khaled patrolled. The marsh on the Iranian Hor al-Azim was initially fed by water from the Karkheh River in Iran. Multiple dams have now been constructed on the Karkheh, which decreased the amount of water reaching Hor al-Azim. Coupled with the new embankment, the flow of water from Hor al-Azim into Hawizeh has been reduced to a trickle. Some years it ceases altogether.

Historically, Hawizeh and Hor al-Azim were a single marsh. The people who lived in and around the marsh had no idea where the border was, nor did they care—at least not until the Iran–Iraq War of the 1980s. And although things returned to normal for a while after the war ended, everything changed when Iran built the dike, effectively putting a physical barrier along the border and preventing both water and people from moving back and forth. The dike acts like a dam, with culverts and weirs,[2] or escape valves, built in should the water rise too high on the Iranian side. Hor al-Azim is now the dam's reservoir and water can be emptied, maintained, or released subject to Iran's needs. The implications for Hawizeh are grave. No longer is there any guaranteed recharge coming from Iran; Hawizeh now depends almost entirely on the flow of the Tigris River.

The dike serves four purposes for Iran. First, it allows the country to open more oil concessions in Hor al-Azim. Second, rather than helping replenish the Iraqi Marshes, water can be utilized for agriculture. Third, it acts as a barrier against smugglers. And fourth, it offers a strategic benefit to Iran by denying Iraq the water. Indeed, in 2008, Iran dried out Hor al-Azim for oil development, in the process stopping the flow of water into Hawizeh. Oil fields underly both the north and the south of Hor al-Azim. In Iran, like its neighbor, oil is king.

Problems arose soon after the marsh was drained. The absence of any vegetation meant that winds could easily pick up and transport the dry soil, and soon major dust storms blanketed the Iranian city of Ahwaz. After protests over both dust storms and the loss of ecological habitat, Iran decided to reflood 80 percent of Hor al-Azim. This allowed some water to return to Hawizeh during wet years, when water flowed through the culverts built high in the Iranian embankment. However, water from the Karkheh is less important to Iraq during wet years since the Tigris provides ample water for the Marshes. In dry years, however, like in 2009, it intensified the impact of drought.

Jassim visited Khaled in early September 2009, accompanied by a press and media team from the al-Fayhaa satellite TV channel. The crew hoped to produce a film about Iran's exacting some measure of revenge for the Iran–Iraq War and other past grievances by building a dike along the border, which would also impact the region's natural environment and its settlements. Khaled served the visitors tea and spoke about the dike and the Marshes. He also made it clear that he was under strict orders from his superiors not to allow filming of the border area without prior authorization.

Jassim was undeterred. The film crew had traveled all the way from Basra for the story, and Jassim decided to make a deal with Khaled. If his friend allowed them to create the video, Jassim guaranteed he would not release the film until he personally met with Khaled's supervisor in Basra and obtained permission. Acknowledging that the story of Hawizeh was too important to ignore, Khaled requested a military truck and asked the driver to prepare it for a trip to the border. When the vehicle arrived, Jassim, Khaled, and the media representatives boarded, with the stipulation that cameras be hidden from view. The Iranian guards in the towers were likely to fire on anyone attempting to film the dike.

Driving through a dust storm, they proceeded to an old abandoned Iranian guard tower where they filmed the devastation of Hor al-Azim. Khaled knew he risked punishment for allowing the filming or, worse, one of the group members might be shot by the Iranians. But Khaled decided he needed to defend the Marshes—he had seen the devastation caused by war, draining, and dams. There was also the important story of the defectors and revolutionaries trying to escape the Iraqi regime, who were sheltered by local villagers. Their stories needed to be told.

In the end, Jassim accompanied the film crew back to Basra and convinced Khaled's supervisor of the importance of the documentary. Khaled received no disciplinary action, and the film proved to be so popular that the station played it ten times over the next few months.

Iran's construction of the dike in Hawizeh, along with new dams on rivers that previously fed the Marshes, had deleterious effects on Iraq. What had been a single ecosystem was now divided in two, and the wetlands on both sides of the border suffered. Hawizeh and Hor al-Azim are—or rather were—interdependent. Their ultimate survival required that both countries cooperate. Acknowledging the need for joint management, the

Ramsar Convention secretariat sponsored a meeting in 2019 with wetlands experts from both countries and UN participants, hoping to convince the countries to collaborate. As one of the leading experts on the Marshes, Jassim attended the meeting; however, little was accomplished. The Iranians were unwilling to talk about water and have since blocked the release of a report on the meeting.

Thus, the dike along the border continues to impact Hawizeh. Although there were technical discussions between Iran and Iraq over Hawizeh Marsh up until 2019, these have ceased. As does the Turkish government, the government of Iran believes it has absolute sovereignty over the water within its boundaries, with no obligation to share with other countries.

The latter half of the twentieth century could be described globally as the century of dams. They sprouted up across the landscape like gangly weeds in an unplowed field. Mesopotamia, with a long history of controlling what little water was available, was not immune from the tsunami of construction—particularly in the latter half of the century. Construction continues unabated today (figure 28). At the very bottom of the two meandering rivers lie the Iraqi Marshes. Any upstream activities affect both the quantity and quality of water reaching the Marshes and, ultimately, the communities dependent on the wetlands.

Kemal Atatürk, the father of modern Turkey, had ideas about diverting water from the Euphrates River for irrigation in the 1930s. It wasn't until the 1960s that initial studies were conducted on building dams and diverting the river for both irrigation and hydroelectric purposes. The concept of fostering greater agricultural development while supplementing the country's electricity supply was too appealing to ignore. By the 1970s, plans were in place for a Southeast Anatolia Project or Güneydoğu Anadolu Projesi (GAP) to advance regional development in southeast Turkey. When completed, it would include eighty dams and sixty-six hydropower stations on the Tigris and Euphrates Rivers. Syria and Iraq were understandably worried.

In the meantime, Syria was building its own massive dam, the Tabqa Dam on the Euphrates River, roughly forty kilometers north of the city of Raqqa. The sixty-meter-high dam was designed to produce electricity and provide irrigation water for agriculture. In the process, it created Lake Assad, Syria's largest water reservoir. When Syria began filling the reservoir behind the dam in 1974, Iraq complained that the flow of the Euphrates had dropped to 25 percent of normal. It didn't help that Turkey was simultaneously filling the reservoir behind the Keban Dam, also on the Euphrates, or that there was a significant drought throughout the region. The diminished flow of water into Iraq drastically reduced agricultural production. The Iraqi government threatened to bomb the Tabqa Dam, and both Syria and Iraq sent troops to the border. In 1975, Saudi Arabia and Russia were brought in to mediate, and eventually an agreement was signed ensuring that 58 percent of the flow of the Euphrates at the Turkey–Syria border would flow into Iraq.

Cooperation over water in the Euphrates and Tigris River basins dates to the dissolution of the Ottoman Empire in the early twentieth century. Britain and France included references to coordination and sharing waters in various conventions and agreements between Iraq and Syria in the 1920s. The first legal instrument of cooperation was the Treaty of Friendship and Good Neighborly Relations between Iraq and Turkey in 1946, obligating both parties to share water flow data and to consult one another on water development projects.

Iraq has enormous reserves of oil. Turkey, on the other hand, has none. Using hydroelectric power to offset at least some of Turkey's energy deficiencies was thus an appealing option for the country. The success of the Keban Dam spurred Turkey to consider a larger regional development project in the southeast, involving dams on both the Euphrates and Tigris Rivers. For the last half-century, the policy of the Turkish government has been that they have absolute sovereignty over water resources that originate in their country. In effect, they own the water in the Tigris and Euphrates Rivers within their boundaries and can do with it as they wish. Sustainable energy, water for agriculture, and economic development in an impoverished region, in addition to fostering internal security, all contributed to the decision to pursue the GAP project. The magnitude of the GAP project and its twenty-two major dams raised alarms in downstream regions where river flow would undoubtedly decrease. Turkey did not care.

The GAP project officially began in 1977, with early construction focused on the Euphrates River. Tensions again rose among the riparian countries. Partly in response to the project, Syria began supporting the Kurdish Workers Party (PKK), which aimed to create a Kurdish state

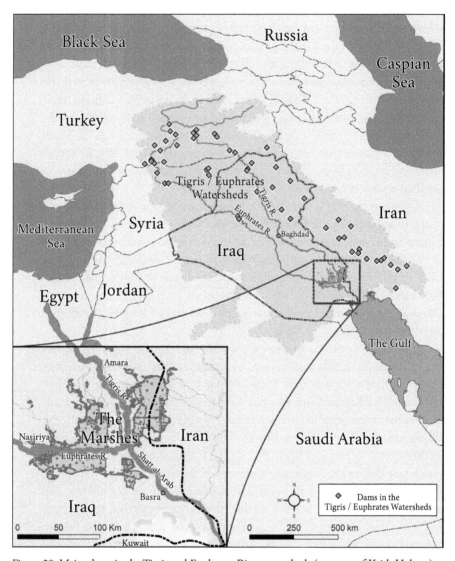

Figure 28. Major dams in the Tigris and Euphrates River watersheds (courtesy of Keith Holmes).

in southeastern Turkey. In 1987, to reduce tensions, Turkey and Syria signed a Protocol on Economic Cooperation which stated that Turkey would provide a minimum flow of 500 m³/second on the Euphrates where it enters Syria if the country reduced its support for the PKK. The agreement did not last.

Turkey continued to complain about Syrian support for the Kurds, informing both Syria and Iraq in 1989 that it would be stopping the flow of the Euphrates River for at least one month—ostensibly to fill up the reservoir behind the new Atatürk Dam, one of the largest in the world. It also signaled to downstream riparians that the country was willing to use water for strategic purposes—an announcement that brought international condemnation on Turkey. Water was the one issue that could make Syria and Iraq cooperate, and in 1990 the two countries signed the Syrian–Iraqi Water Accord. It guaranteed that a minimum of 58 percent of Euphrates River water entering Syria would flow across the border into Iraq, much like their initial agreement in 1975. This meant that Iraq would receive a minimum flow of 280 m^3/sec at its border with Syria—provided Turkey met its commitments.

The Euphrates River, the driver of economic development in southern Mesopotamia for thousands of years and the lifeblood of the Marshes, stopped flowing into Syria and Iraq in January 1990, when Turkey began filling the reservoir behind the Atatürk Dam. The response by Syria and Iraq was swift, with both planning armed retaliation against Turkey. They also boycotted all companies involved with the GAP project. Turkey remained resolute. The hope was to fill the reservoir in two months. In early March, with the reservoir less than half full and international pressure mounting, Turkey agreed to release water downstream and meet their commitment to Syria. Water soon began flowing into Iraq as well. In the end, it took two years to fill the reservoir behind the Atatürk Dam. For the moment, conflict was averted.

The plan was for the GAP project to be completed by 2010; however, financial constraints have delayed this to 2047. Roughly half of the planned projects have been implemented thus far.

Dam construction on the Tigris and Euphrates Rivers has had two noticeable long-term impacts on Iraq. First, the variability of water flow month to month decreased; no longer were there seasonal differences in flow, or peak floods that can wreak havoc on villages and agriculture lands. This has been beneficial for both Syria and parts of Iraq. On the other hand, the natural overflow from the rivers during winter months is what traditionally replenished and cleansed the Marshes. The fresh water diluted the existing water in the Marshes and assisted in pushing some of it out of the system—nature's elegant way of providing fresh water for plants and animals. Today, this only occurs in very wet years.

The second impact of the dams on Iraq was a reduction in the overall flow of both the Tigris and Euphrates Rivers. People in the Marshes do not need to look at the data; they know the rivers are saltier and more polluted than before, and that water levels, particularly on the Euphrates, are lower than they were before the draining of the early 1990s. Added to this have been four major droughts in the past fifteen years that negatively affected nearly all activities related to the Marshes.

Arguably the most controversial dam in Turkey was the Ilisu Dam. The reservoir for Ilisu began flooding in late 2019. By mid-2020, 200 villages had been flooded, including the ancient city of Hasankeyf, and 80,000 people were displaced.[3] Biodiversity will likely be affected, with the expected loss of at least four endangered animal species. The impacts downstream in Iraq include much reduced water flow which, in turn, affects both water availability and power production; higher levels of salts and other pollutants; and the loss of water to help replenish the Marshes. Nature Iraq, which organized the flotilla of boats down the Tigris River, was one of many groups clamoring for a reconsideration of the Ilisu project. Jassim was now working for Nature Iraq, and he understood the potential negative impacts of Turkey's dams on the Marshes.

Turkey is now building a fifth dam on the Tigris River near the city of Cizre, downstream from the Ilisu Dam and just forty kilometers from the border with Syria. While the Cizre Dam is smaller than Ilisu in terms of hydroelectric production, it will have a greater impact on the Marshes. The dam captures water released by the Ilisu Dam and then diverts it for irrigation purposes. This, in turn, will further reduce the flow downstream. In effect, Turkey now controls the flow of both the Tigris and Euphrates Rivers, leaving the Marshes particularly vulnerable.

Jassim and Nature Iraq, along with their flotilla of vessels that included the guffa, the kalak, and the tarada, were unsuccessful in stopping the building of the Ilisu Dam and the flooding of Hasankeyf. No amount of public pressure, whether national or international, would distract Turkey from achieving its goal of damming the Tigris River. Nevertheless, the project inadvertently raised awareness of the threats to the river and its biodiversity, particularly as it flows into the Marshes. Yet whether greater public awareness would influence government policy in Iraq was uncertain. The Ministry of Water Resources in Iraq criticized the Nature Iraq effort, maintaining it could jeopardize ongoing negotiations with Turkey over water-sharing agreements. In the meantime, the actions of another neighbor were affecting the one marsh that had survived draining by the Ba'th government two decades before.

Jassim grew up where civilization began and has spent a lot of time thinking about the relationship between the Marshes and the once great cities of Ur, Uruk, and Eridu. One of his friends is Dr. Jane Moon, an archeologist from the United Kingdom whose research focuses on the ancient area of Tell Khaybar, near the city of Ur. Jassim accompanied her many times on trips to the Marshes and nearby archeological sites. Dr. Moon shares Jassim's belief that every culture defines its relationship with the ground under its feet and the plants and animals that spring from it. It is part of the essence of being human.

Franco D'Agostino, an Italian professor of archeology who specializes in Sumerian philology, is another acquaintance of Jassim's. D'Agostino's work focuses on Abu Tabira, a Sumerian site seven kilometers south of Nasiriya whose residents engaged in fishing, animal husbandry, and weaving reed mats, much like the Ma'dan in the latter half of the twentieth century. Jassim often joins him on trips to Abu Tabira to discover more about the history of human presence in the Marshes.

These friendships, along with Jassim's extensive network of contacts in government and villages throughout the region, as well as his knowledge of the wetlands, have helped him emerge as an effective advocate for the Marshes. He is sought out by journalists and film crews who want to report on the Marshes, by diplomats and high-level officials who want to see for themselves what this unique ecosystem is like, and even by artists hoping to paint and photograph the region. On occasion, these activities cause him trouble as well.

In March 2017, Jassim was invited to escort Barham Salih, future president of Iraq, and 'Adil 'Abd al-Mahdi, soon to be prime minister, on a tour of the Central Marsh. They rode in a fiberglass shakhtura from the north bank of the Euphrates River in Chibayish to Baghdadiya Lake in the middle of the marsh. It was a relaxing day for all, giving Jassim an opportunity to speak to senior politicians on the history and ecology of the Marshes, the activities of the local people, and the ways in which the Marshes could be preserved. At the end of the trip, there was the requisite photo with Jassim, Salih, and Abdul Mahdi, which was posted on digital media the following day.

A year later, shortly before the federal election that would put both Salih and Abdul Mahdi in power, the photo was reposted by someone whom Jassim had never heard of, with the caption, "The American Intelligence Officer for Middle East Affairs, Richard Funk, with Barham Salih and Abdul Mahdi in a mash-huf in the al-Chibayish Marshes. Oh, how big you are Uncle Sam, how naïve you are, with your purple finger."[4]

In a digital instant, Jassim went from being an irrigation engineer and director in Nature Iraq to an American spy trying to influence the upcoming Iraqi elections. Nevertheless, he was fortunate to have friends from around the world come to his aid. The response was overwhelming as people not only defended him but praised his work and his connections to both scientists and local residents. One comment that went viral was from Dr. Ahmed Amin, a physician and well-known human rights activist. "The fruitful trees are always stoned. It is not possible to deny your efforts and what you have done for the sake of our homeland. May God guide you along your way."[5]

Dr. Mudher Muhammad Salih, an economist and financial advisor to future Prime Minister Mustafa al-Kadhimi, went further in his October 2018 post to a website on civil disobedience. The piece was titled: "Is the Blond Iraqi a Curse or a Blessing?"[6] In the article, Salih recounts the story told to him by his father about Samim al-Saffar, a tall, white, blond man from Baghdad who spoke perfect English. Al-Saffar was a volunteer doctor helping Iraqi troops in Fallujah in 1941 during the Battle of Sin al-Dhuban in the Anglo–Iraqi War. Salih's father was a civil servant in Fallujah at the same time and watched al-Saffar treat patients as the battle escalated. Just prior to the evacuation of the city, al-Saffar was treating civilians injured in the fighting when a group of Iraqi soldiers surrounded him, cried out that he was a British spy, and killed him. Salih told his son that it was no wonder Iraq lost the war with a fighting force that could not distinguish a British soldier from a volunteer Iraqi doctor.

Muhammad Salih ended his article by comparing the story of Samim al-Saffar to the claim that Jassim was a U.S. operative, noting that Iraq's confused history in its fight against colonialism hadn't changed in seventy-five years. In fact, there was universal condemnation of the photo with the caption naming Jassim as an American operative—with one exception. He received a call from the leader of a local militia asking him whether he was, indeed, working for the CIA in the U.S. If so, the caller claimed, he would be killed. Jassim convinced the caller that he was not Richard Funk and had no connection with the U.S. government. The caller was quiet for a moment and then hung up, apparently satisfied with Jassim's explanation. The bizarre case of mistaken identity was covered extensively by television, newspapers, and the digital media. Unexpectedly, it gave Jassim the public exposure he needed to become an important voice in favor of the Marshes and against those who would prefer development over preservation. He now had an audience, and he made the most of it.

Jassim's interest in archeology gave him a strong sense of the history of the land that he so revered. Working with Nature Iraq also gave him a better sense of the importance of biodiversity to a nation's identity and development. The environmental organization, in collaboration with federal and regional ministries and Iraqi academics, conducted an intensive study of key biodiversity areas of Iraq from 2006 to 2011. Eighty-two sites around the country were surveyed multiple times in summer and winter to provide an inventory of Iraq's biodiversity. The study was invaluable in supporting the proposal for the Mesopotamian Marshlands National Park and the UNESCO World Heritage site submission. The problem was finding someone to fund the publication of the study, and it took several years—and some unexpected publicity—for Jassim to find the funds.

Not long after his excursion with Salih and Abdul Mahdi, Jassim was invited to a meeting in Basra with senior members of the Ministries of Environment, Oil, and Water Resources. The topics ranged from preserving the Marshes, to pollution from oil development. Jassim noted the Oil Ministry's growing interest in biodiversity issues, and he proposed it help publish the study. His pitch succeeded.

Six years after the study was completed, "Key Biodiversity Areas of Iraq" was published to international acclaim. The director general of the World Wildlife Fund noted, "This work unveils the extraordinary natural richness of Iraq, something the Iraqi people should be proud of, and the international community should help preserve. It is a great contribution to fulfilling what is our most existential aspiration and challenge: live in harmony with Nature."[7]

Chibayish remained cut off from the rest of Iraq until the late 1960s. Life was simple, and residents lived either in town or on one of the many islands that dotted the Central Marsh to the north and Hammar Marsh to the south. The residents had the sustenance they needed, with ample fish, birds, beef, eggs, and chicken. The one dirt road that connected Chibayish with Nasiriya—100 kilometers to the northwest—was little more than a windy, bumpy track and it was a tortuous ride by truck or car. Few Iraqis ventured all the way to Chibayish, and even fewer international visitors. It was therefore quite a surprise to the local population when a group of white-skinned, blue-eyed visitors arrived one day and set up a camp near the village of al-Khater in Chibayish District. More perplexing was the collection of amphibious vehicles that accompanied them, along with various types of equipment. It caused quite a stir over coffee in the main mudhif on the outskirts of the village: Who were these foreigners? What was their interest in the Marshes?

Fadi al-Khayoun understood what was happening. He was a friend of Jassim's father and one of the few residents who knew about life outside the Marshes. He told the men gathered in the mudhif for coffee that the visitors were French and were looking for oil. One of the men asked Fadi what he meant by the word "French."

Fadi laughed loudly. "I mean they are from France."

Fadi's knowledge of France wasn't born from his travels, but from his drinking. He enjoyed classic French wine from Bordeaux. His worldly knowledge came through his love for alcohol, which was unusual in a culture and religion where consumption of alcoholic beverages is uncommon.[8] When another man asked about the vehicles with very large tires, Fadi explained that these were special vehicles that could walk through the Marshes and make their own waterways in search of oil. The questions kept coming and Fadi responded confidently, though often his answers were contrived.

"Do the French drink alcohol?" one asked.

Fadi told the gathering that the visitors were sons of the Prophet Jesus, and Christianity did not prohibit them from drinking alcohol.

Another asked why the Prophet Muhammad forbade Muslims from drinking, and the mudhif went quiet. Fadi was embarrassed at the question and thought about it carefully. After a few moments, he told them that God revealed to Muhammad that his followers should not drink alcohol because it causes them to lose faith. Raysan interrupted Fadi and asked him why God did not tell Jesus to ban alcohol, as he did with Muhammad.

Fadi replied, "God commanded Jesus to prevent his followers from drinking alcohol, but Jesus said, 'My Lord, my followers cannot do that. They love fun, joy, singing, and drinking.' So, the Lord said, 'You will have what you want.'"

This explanation brought a round of laughter from the assembled group. One fellow even challenged Fadi to show the group where this was written.

The following day, the residents of al-Khater village were surprised to see Fadi working as a guard at the French camp. A few of the men were quite certain that the French chose Fadi as a guard because he drank alcohol, just like they did. Over the next few weeks, twenty-five others were employed by the French to assist them in exploring for oil. People were fascinated by the amphibious machines and the use of explosives to facilitate seismic surveys of the area, and many paddled their boats behind the vehicles to observe what was happening.

Wine and oil weren't the only products in which the French were interested. News spread that the French workers enjoyed hunting wild boar; they would cook their catch over an open fire and enjoy a feast. The Central Marsh was full of pigs, and although none of the locals ate pork, a few were more than willing to hunt boar and sell them to the French.

After a year of exploration, the French company left the area without finding any evidence of oil under this section of marsh—most of the discoveries were to the east and south of Chibayish. As for Fadi, he enjoyed a year of French cuisine, whiskey, and hunting duck and boar with the visitors.

The Majnoon oil field in southern Hawizeh Marsh contains over five billion barrels of oil and is one of the largest in the world. Jassim and his colleagues had little sense of this when they worked to build embankments in the region to aid Iraqi forces during the Iran–Iraq War. Twenty-five years later, Jassim returned to Majnoon, this time as part of a team that was assigned the task of assessing the social, economic, and environmental impacts of oil operations. The work included evaluating impacts on biodiversity and local communities. When Jassim arrived, he was startled to see that the area was devoid of life. Not only were the Marshes gone, but there was no agriculture, despite the Ministry of Agriculture establishing the Safia Reserve in 2004 near Majnoon to promote biodiversity, with an emphasis on local fish. There was nothing but dry and barren land.

Iraq is awash in oil. It has the fourth-highest oil reserves in the world and is ranked sixth in the world for total production. The country is an important member of the Organization of Petroleum Exporting Countries (OPEC). OPEC was formed in 1960 by Iran, Iraq, Kuwait, Saudi Arabia, and Venezuela to stabilize—although many would say *influence*—oil prices. OPEC now includes thirteen countries with over 80 percent of the world's oil reserves and 45 percent of global production. Although OPEC's power has waned in recent years as production increased elsewhere, it continues to have a significant influence on the price of oil through production quotas agreed to by its members.

The problem for Iraq and other energy-producing countries is that the price of oil is inherently unstable, despite attempts to control the supply. In 1998, the price of a barrel of oil was below $20. Ten years later, it was over $150. By 2020, it had again dropped to under $20 per barrel. During the spring and summer of 2021, the price stabilized in the range of $60– $70 per barrel. For a country such as Iraq which is almost entirely dependent on oil revenue, these fluctuations make the economy quite vulnerable.

Oil projects in Iraq involve a system of government-awarded servicing contracts and licensing agreements with international companies to manage the oil fields. Iraq retains a minimum 25 percent share in all fields. The resulting revenue from oil production accounts for over 99 percent of Iraq's foreign-exchange earnings and over 90 percent of all government income. Ninety-seven percent of the country's electricity is produced using oil or natural gas, with the remainder being hydroelectricity. Oil drove economic modernization in Iraq during the 1960s and 1970s, and it allowed the country to engage in costly conflicts with Iran and Kuwait, and the U.S. oil does not drive the economy; it *is* the economy. More importantly, many of Iraq's oil fields lie in, or near, the Marshes (figure 29).

Iraq's large oil reserves should make the country and its citizens very wealthy. Instead, Iraq suffers from high levels of poverty, is in the lower third of all countries in human development (a measure of social and economic well-being), and is mired in debt. It must import gasoline for vehicle use, and over one-quarter of the country's electricity is either imported directly or produced with imported natural gas. The singular reliance on oil places Iraq in the bottom quarter of the world in terms of economic complexity, which is a measure of economic diversity.[9] Rather than using oil revenues to diversify its economy and enhance its citizens'

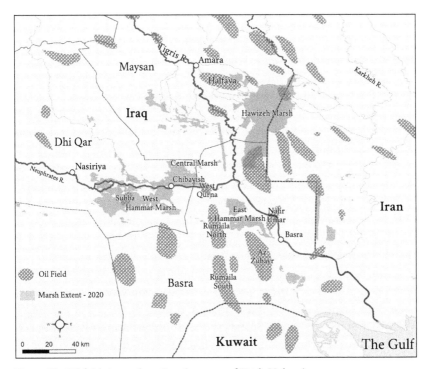

Figure 29. Oil fields in southern Iraq (courtesy of Keith Holmes).

well-being, Iraq's dependence on oil has increased in the past forty years. At present, monthly revenues from oil do not cover the salaries of all the civil servants in the country. However, this is not to say that there have been no improvements in the standard of living for Iraqi citizens, including those living in or near the Marshes. Indeed, services such as healthcare and education have improved—but at a cost.

Oil production in Iraq is concentrated primarily in two regions: the governorates of Kirkuk, Erbil, Dohuk, and Sulaymaniyah in the north; and Basra, Maysan, and Dhi Qar in the south. Two of the largest oil fields in the world are the Majnoon field in Hawizeh Marsh and the North Rumaila field in Eastern Hammar Marsh—or rather, they were in the Marshes until these areas were drained for oil development. Majnoon and Rumaila are supergiant oil fields, meaning each contains over five billion barrels of oil. Two other supergiant oilfields, West Qurna 1 and West Qurna 2, lie in former sections of the Central Marsh and East Hammar Marsh. Crude oil from the Iraqi fields is pumped into a main pipe that parallels the Tigris and Shatt al-Arab Rivers and moves south to the Gulf, where it is trans-shipped via tanker around the world.

The Majnoon field, which became famous when Iran occupied the site during the Iran–Iraq War, was discovered in the Hawizeh Marsh in 1975. There are plans to double its output by 2030, with more drilling north into the Marsh. A second large development, the Halfaya oil field, abuts the UNESCO buffer zone around Hawizeh in the northwest. Halfaya was discovered in 1976 but began operating only in 2010. It, too, is expected to double output in the next ten years. There is a third, non-producing oil field in the north of Hawizeh, known as the Hawaiza field. Only one-fifth the size of Halfaya, it not only lies inside the Marsh, but is adjacent to Umm al-Ni'aj, the largest open-water lake in the Marshes. Test wells have been drilled in the Hawaiza field, but no date has been set for production to begin.

During preparation of the Iraqi submission to UNESCO in 2014, oil development in the Marshes proved to be an important and concerning issue. Not surprisingly, oil interests in Iraq were worried that creating a World Heritage site in the Marshes would inhibit future oil field development. On the other hand, the UNESCO committee believed that oil development might compromise the integrity of a future World Heritage site. The Iraqi committee preparing the submission worked closely with the Ministry of Oil to ensure that any future World Heritage site would not overlap with existing or future oil fields, and that there would be at least a five-kilometer buffer zone between any natural or cultural site and an oil field. This, however, may not have been enough. The vulnerability of aging pipelines, the ongoing possibility of terrorist attacks, the discovery of new fields and expansion of existing ones, and the need for water to extract oil all threaten the wetland ecosystem. Still, the impacts of oil activities can be minimized through slant drilling, strengthening petroleum laws and procedures related to drilling, improving extraction and transportation infrastructure, and better adhering to environmental standards for the working and preserving of the hydrological system of the marsh waters.

Oil production in Iraq, while not fluctuating as much as oil prices, has been variable since 1980 due to wars and international sanctions. Production dropped precipitously at the start of the Iran–Iraq War, as Iran either bombed or captured strategic oil sites in Iraq. Transportation of oil to international markets was also jeopardized when Syria supported Iran in the war and blocked the pipeline that transported oil from Iraq to tankers in the eastern Mediterranean Sea. Losses in the oil sector and the accumulation of foreign debt crippled the country.

Production rose steadily after 1983, and by the time Iraq invaded Kuwait in 1991 it had again reached 1980 levels. Saddam had hopes of using Kuwaiti oil to support both his military expenditures and economic modernization program. Instead, he was confronted with extensive damage to infrastructure from coalition forces as Iraq was driven out of Kuwait during the Gulf War. Subsequent international sanctions on the country drove production to near zero through the 1990s, taking an enormous toll on Iraqi citizens. The loss in revenue reduced government services, and inflation soared.

Despite civil unrest in the country since 2003, and hundreds of bomb attacks on oil facilities and pipelines, oil production has rebounded. In early 2020, Iraq produced over four million barrels per day—higher than its two previous peak years of 1980 and 1990 (OPEC production cuts have since lowered this figure). However, the economy remains in dismal shape, with high inflation and low demand. Economic growth in 2020 fell by 11 percent, partially due to the COVID-19 pandemic. Iraq's level of debt relative to its GDP is presently the highest of any OPEC country. For the first time in almost twenty years, the dinar was devalued, and the government was forced to draw on its financial reserves to pay salaries.

The three marsh governorates of Dhi Qar, Maysan, and Basra have been among the hardest hit economically: unemployment is high, and one-third of residents are below the poverty line. The economic situation is reminiscent of the late 1990s, with the government unable to cover the wages of its employees. Nevertheless, oil will continue to drive Iraq's economy through the twenty-first century, even as global demand for oil slows due to policies aimed at reducing greenhouse gas emissions. Oil talks and the government listens.

The abundance of oil in Iraq, however, is in stark contrast to the electricity supply. Most homes receive power for only twelve hours per day, on an alternating system of two hours on and two hours off. This approach is necessary because peak demand exceeds Iraq's installed electrical capacity by 50 percent. The actual amount of electricity available is much lower due to inefficient and damaged power plants and poor transmission infrastructure. Diesel generators are almost a necessity. In fact, electricity production is so inadequate that Iraq now imports electricity and natural gas from Iran, with the U.S. granting Iraq a special sanctions waiver to its policy of prohibiting any country from purchasing Iranian energy. In 2017, Iraq began importing natural gas from Iran to fuel electric power plants near Baghdad. Twenty-three percent of all electricity in Iraq is generated with gas produced in Iran, and 5 percent is imported directly. And although these imports are essential supports for a crippled electrical system, the relationship between both counties may not last long. For one thing, Iraq is over five billion dollars in arears to Iran, and there have been threats to cut off the exports for non-payment. Iran, too, is facing economic difficulties, owing to a combination of international sanctions and lower oil prices.

The relationship between oil and water in the Marshes is complex. It starts with the fact that oil production demands a lot of water. It can be fresh water, salt water, or even polluted water (with some exceptions), but there must be water to inject back into wells to replace lost oil and maintain reservoir pressure. If one million barrels of oil are extracted from a well, then one and a half times as much water must be injected. There are four possible sources for the injected water. Some comes from the water pumping station in the village of Abu Sakhir near Basra Airport, which treats a large quantity of water from the Shatt al-Basra Canal and distributes it to the oil fields in Basra and Maysan. Another source is the Main Outfall Drain (M.O.D.) Canal before it empties into the Shatt al-Basra. The M.O.D. water is de-oxygenated—which is necessary to prevent corrosion in pipes and bacterial growth in wells—and then injected into the ground. It makes sense to use water that is polluted with salt and is heading for discharge into the Gulf as injected water in the oil fields, as opposed to valuable fresh water. The M.O.D. is carefully monitored to ensure sufficient water to maintain a continuous flow into the Shatt al-Basra Canal and out to the Gulf. There isn't enough water in the M.O.D. to satisfy all the needs of nearby oil fields, but it helps.

A third source involves re-injecting water produced in the drilling process back into the wells. When oil is pumped out of the ground, it is accompanied by produced water. As much as two-thirds of the liquid that comes out of the ground is a combination of water and dissolved solids. Prior to transportation by pipeline, the oil needs to be separated from this polluted water. There are a few different processes for this, but the least expensive one, assuming land is available, is to emulsify the mixture in large settling ponds. The oil rises to the top of the mixture—much as oil-and-vinegar salad dressing separates after a few minutes—and the oil on top can then be removed. The water is treated, and any sludge is removed before it is re-injected into the wells. Recycling the produced water can contribute 60 to 70 percent of the water a well needs.

A fourth possible source for the injected water is from treated seawater. In 2011, the Iraqi Ministry of Oil, in collaboration with Exxon, announced the Common Seawater Supply Project (CSSP), an ambitious plan to take seawater from the Gulf, treat it to remove some of the salt, and then pipe it to oil projects in southern Iraq. Over 400 kilometers of pipeline would deliver 13 million barrels of treated seawater per day to the oil fields for injection. Exxon walked away from the project in 2018 when negotiations over the contract broke. This was due in part to the corrupt business environment that is pervasive in the oil sector in Iraq.[10] Although a recent agreement with a French company may revive the CSSP, no one is speculating about a completion date for the project.

A combination of produced water and polluted water from the M.O.D. and Shatt al-Basra Canal provides roughly 85 percent of the water needed for injection. The only other sources of water are nearby rivers, canals, and marshes. The Ministry of Oil will not divulge exactly where the additional water comes from, even to other government ministries. For instance, CRIMW and the Ministry of Water Resources were unable to obtain Ministry of Oil data on water use in the oil sector for their 2014 study on the impact of the oil industry on the Marshes.

The expansion of oil development and the use of water from rivers and marshes to inject into wells are two matters of contention between oil interests and Marsh preservationists. The CRIMW–Nature Iraq study on the impacts of oil development in the Marshes highlighted two additional concerns. The first relates to the disposal or leakage of produced water that emerges with the oil during drilling. There is potential for leakage from the large holding ponds used to emulsify the extracted oil, but to date there has been no evidence that this has affected the Marshes. If the integrity of the dikes and ponds is maintained, the Marshes should not be impacted.

The second issue concerns the problem of pipeline infrastructure. Pipelines can safely transport liquids such as oil, with two caveats. The first is that the likelihood of a pipeline failure or leak increases with age.

There is anecdotal evidence of small leaks in pipelines in the Marshes, though neither the government nor industry has acknowledged this. The second very real possibility is through deliberate sabotage of pipelines, which have become targets of insurgents and terrorists, and Iraq has been particularly hard hit. Between 2003 and 2008, over 300 incidents involving pipelines and storage facilities were reported.[11] Most of the attacks occurred in the northern part of the country, but the south has not been immune. As recently as November 2020, a gas pipeline exploded in southern Iraq, although no group has claimed responsibility. Although attacks on pipelines and storage units have decreased markedly in recent years, these facilities remain vulnerable—even a small spill in the Marshes would have a significant impact on the ecosystem.

Little impedes oil production in Iraq. Oil revenues are vitally important to the economy, and they will continue to play a crucial role for at least the remainder of the century. Having major oil production facilities in or near the Marshes is not ideal for wetlands preservation. On the other hand, international oil companies today, including those active in the Marshes, are much more concerned about their environmental and social footprints than they were three decades ago. They are subject not only to national policies and constraints, but must adhere to international environmental norms and standards, regardless of where the oil is produced. While these companies may have much to atone for, new safety standards and options for injected water may reduce impacts on the Marshes. The Iraqi Ministry of Oil is aware of the potential conflicts and has worked with the Iraqi committee on obtaining World Heritage site status for the Marshes. The Common Seawater Supply Project is another example of the government and the private sector trying to find innovative alternatives to using fresh water in the production process. The Ministry of Oil and the Basra Oil Company also supported the publication of Nature Iraq's book *Key Biodiversity Areas of Iraq*.[12] These are all positive signs.

Nevertheless, oil production never provided the economic benefits that were promised to marsh communities, nor did it replace the activities that were lost. There were two reasons for this: first, the number of marsh dwellers working in the oil fields is quite small since workers with technical expertise are brought in from outside the region; and second, the promise from the government that 5 percent of the value of oil production in the Marshes would be designated to building Marsh villages was never realized. Politicians and community leaders redirected most of the funds to cities to gain electoral support. Still, these problems, whether economic or environmental, can be rectified. There is no alternative but for interested parties to cooperate since both oil production and Marsh preservation are essential to Iraq's future.

8

A PLAGUE OF DOUBT, A PATTERN OF DROUGHT, AND A REASON FOR HOPE

Al-Ruta was one of many neighboring villages located in the southwest section of Hawizeh Marsh. Reed huts in the various villages created an elegant symmetry when viewed from above as they formed a winding path among the patchwork of water channels that stretched to the east, toward the Iranian border. Families living in al-Ruta harvested reeds, fished, and tended buffalo, as befitted Ma'dan living throughout the Marshes. The same had been true for a thousand years or more. Umm Hussein was born in al-Ruta in 1958 and spent her formative years there. Her life in the village was not much different from that of Najmah's, except that Umm Hussein had no formal schooling.

When Umm Hussein came of age, she fell in love with Abu Hussein. On her wedding day, women wore bright clothes and danced in joy, boats were loaded with men shooting guns toward the sky, and children were running everywhere enjoying sweets. Umm Hussein's new husband sat bemused in his long, white dishdasha as women ululated all around. For Umm Hussein, it was a wondrous day.

Her marriage was a life Umm Hussein dreamed about as a child. She bore one son and three daughters in al-Ruta before the Iran–Iraq War intervened and changed her life in an instant.

In 1983, the Iraqi army burned down al-Ruta and forcibly removed the residents. Almost every village in the area suffered the same fate. Umm Hussein and her family were first taken to a camp in northern Amara, then to another near Basra, then to three other locations—all within seven years. The Iran–Iraq War ended in 1988, but their nightmare did not. Forced detainees from Hawizeh were not allowed to return until 1993.

By that time, the Iraqi government had implemented its plan to drain the Marshes, and al-Ruta was one of the villages that suffered from lack of water. Umm Hussein became one of thousands of internally displaced persons who were forced from their homes and forgotten. Despite the hardships, however, she gave birth to two more daughters—although she was still without a permanent home.

When the earthen embankments built by the Iraqi government to drain the Marshes were toppled in 2003, water began flowing back into the section of Hawizeh Marsh near al-Ruta. Umm Hussein and Abu Hussein, along with their six children, returned to help rebuild the village. But much had changed. Major excavation during the Iran–Iraq War had altered the landscape; there was not as much water as before, and the water that did return was laden with salts from agricultural runoff. The returning villagers faced other predicaments as well. Landmines had been planted throughout the region during the war, and as a result many children were martyred in the early months after Umm Hussein's return. Nevertheless, the reeds, birds, and fish eventually returned, and the family was able to raise a few buffalo. It seemed their life was slowly returning to normal.

Al-Ruta never quite recovered, however. Three years after rebuilding the village, water levels in the adjacent marsh began to fall. Unbeknownst to Umm Hussein and Abu Hussein, Iran was constructing a dike along the border with Iraq that acted as a barricade against water flowing into Hawizeh. Iran also built dams along the Karkheh River which for millennium had emptied into the Marsh. Umm Hussein's dream of returning to al-Ruta and enjoying the life she had as a young child would never be realized—there simply wasn't enough water.

Less water flowing into the southwestern section of Hawizeh Marsh made the residents of al-Ruta more vulnerable to the droughts that occasionally plagued Mesopotamia. The drought of 2009 dried the wetlands around al-Ruta and made it difficult for the residents to make a living. It was as if the war and the draining were happening again. The following year offered some respite, as rains in the north were plentiful and the Tigris River provided enough water to replenish the marsh. Two years later, however, Abu Hussein succumbed to a terminal illness. And although Umm Hussein continued to have the support of her daughters (most of whom had families of their own), and her son—who by then had a family of three wives and three sons—life was difficult.

A second severe drought in 2015 dried out the wetlands around al-Ruta completely. This time, Umm Hussein and her family had little choice but to move elsewhere. Leaving the village where she was born and married and where she gave birth to four children was as difficult in 2015 as it had been in 1983. What's more, her husband was no longer there to help and support her. She gathered a few simple items and moved with her family and three buffalo to Hassjah, a village in the Central Marsh where other families from al-Ruta had relocated—yet another new village for Umm Hussein, but at least she would have friends and family nearby.

The village of Hassjah, with its houses of mud and reeds, straddles a paved road that connects the center of Chibayish District with Medina District. When Umm Hussein arrived in 2015, there was a nearby island that housed thirty families, including a few from al-Ruta. Umm Hussein, with agreement from the local sheikh, decided her new home would be on the island, among her friends. On the day she arrived, residents of the village built a reed house on the island to provide temporary shelter for her family; it was their way of greeting an old friend and new resident. Over the next two days, three more dwellings were constructed, and Umm Hussein and her family were now settled. It was very much like living in Hawizeh, with water, reeds, boats, and buffalo spread out in front of her. The buffalo would leave the island in the morning and wander through narrow channels searching for fodder before returning in the late afternoon.

The drought that affected al-Ruta in 2015 was not confined to southwestern Hawizeh; it affected marsh areas throughout the country. Villages farther from the rivers and lakes suffered the most. Water was plentiful over the next two years, but another extreme drought occurred in 2018, and this one affected all the Central Marsh, including Hassjah. The Marshes dried up, the reeds turned brown, and the buffalo had nothing to eat. Umm Hussein was forced to sell one of her buffalo to buy fodder for the other two. Three dry years in the previous decade, along with reduced flow from upstream dams on the Tigris and Euphrates Rivers, imposed severe economic hardship on southern Iraq and those whose lives depended on the Marshes. When the water returned a year later, everyone rejoiced.

Southern Mesopotamia was never immune to serious drought. It is likely that a lack of water 4,000 years ago dried up the Marshes, shifted the course of the Euphrates River, displaced thousands of people, and destroyed an entire civilization. There is no reason to believe it could not happen again. Indeed, when the Iraqi government drained the Marshes in 1992, it was as if an artificial and very extreme drought had been imposed on the region: the Central and Hammar Marshes completely dried up; thousands of people were displaced; and while the course of the Euphrates River remained the same, new rivers and canals were constructed to facilitate the movement of water through and around the Marshes. Heavy machines took only a decade, between the Iran–Iraq War and the implementation of the drainage scheme, to achieve what had taken hundreds of years to occur naturally. Sections of the Marshes did reflood after 2003, but by 2021 there was less marsh, and water quality in many areas was poor. Upstream dams now pose an additional strain on water supply. The Marshes stand at the precipice.

The frequency of drought in the region is increasing. Umm Hussein experienced this first-hand when drought once again returned to Hassjah in 2021, and as before, the wetlands surrounding her house dried out. She had no idea what caused the Marshes to dry up or why this had occurred

Figure 30. Woman sitting on a *mash-huf* in a dry lake in the Central Marsh, 2015 (courtesy of Jassim Al-Asadi).

three times in the past seven years. Jassim explained to her about Iran's dike, the dams built by Turkey on the Euphrates, and climate change. She had never heard of such things.

"What do you think might happen to you in the future if the drought continues?" Jassim asked.

"God is the most merciful" was Umm Hussein's only reply.

The UN Intergovernmental Panel for Climate Change (IPCC) concluded in 2021 that climate change is not only widespread around the globe, but the impacts are both rapid and intensifying. Climate change will manifest itself in higher temperatures, greater variability of precipitation, and more extreme events, such as drought. The implications for the Marshes are worrisome.

Average annual temperatures in the Tigris and Euphrates river basins are expected to increase by more than two degrees Celsius over the next fifty years, with the greatest increases projected for the highlands in southeast Turkey where the two great rivers originate. Winter temperature increases will likely be three degrees Celsius over present-day temperature in the upper part of the river basins, after having risen over one degree since pre-industrial times. Along with the temperature increases are projections of winter precipitation levels 20 to 30 percent lower than today's. This would result in a reduction of snowpack in the mountains—the main source of water for the rivers—of 50 percent or more by the end of the twenty-first century. The scale of these changes and the impacts that may occur are sometimes difficult to comprehend.

Less water going into the rivers will impact both power production and the availability of water for a variety of other uses, including irrigation and replenishing the Marshes. Warmer temperatures will likely translate into higher rates of evaporation and an increased demand for water. It will make for some difficult choices by both governments and private citizens.

Shamash (Utu) is the Mesopotamian god of the sun who helped Gilgamesh, the King of Uruk, in his quest to reach the Cedar Mountain and kill Humbaba. It seemed as if Shamash exerted his authority over the Marshes during the summer of 2020, when the temperature in Chibayish reached fifty-six degrees Celsius (over 132 degrees Fahrenheit), the second highest temperature ever recorded on Earth. A year later, Canada, a country known mostly for snow and ice, recorded a temperature just shy of fifty degrees Celsius in the western province of British Columbia in the summer of 2021. The climate is changing, and those living in countries most vulnerable will be the first to feel the effects—people like Umm Hussein and her family.

The average projected changes do not, unfortunately, tell the entire story. Climate warming will likely be accompanied by climate weirdness, the concept that there will be an increasing magnitude and frequency of extreme events.[1] Drought is likely to occur more often, for longer periods, and to be more extreme. Nevertheless, the pressing issue for the Marshes is less about the decrease in water availability, which is a given, and more about how Iraq responds to the challenges of a warming climate. Improved water resource management that includes reducing the dependency on water intensive agriculture seems an obvious response, but not necessarily a popular political strategy. As when the drought plagued Sumeria 4,000 years ago and brought about the collapse of a civilization, there may be population displacement from rural areas to the cities. There will be a disproportionate impact on women and the poor as well. Buffalo breeders, fishermen and women, craftspeople who work with reeds, and subsistence farmers will all be negatively impacted. Large dust storms, such as those that occurred when Hor al-Azim in Iran was purposely dried out, will occur more often, resulting in more respiratory illness.

Adding insult to injury, the loss of wetlands will make local climate conditions worse. Not only will it cause dust storms, but dry ground absorbs more solar radiation than do lakes or the ocean; it is why people and buffalo get into the Marshes to cool off. For the buffalo, immersing themselves in the water is essential to maintain their core body temperature. Moreover, dried marshland retains heat and makes the region warmer, as opposed to land covered with vegetation, which absorbs carbon dioxide and has a cooling effect on the surrounding area as it also absorbs, then releases, water. Although the Marshes do not absorb the amount of CO_2 as does a northern temperate forest, the loss of any vegetation means less CO_2 uptake and more heat.

The capacity of the Iraqi government to respond to the stresses posed by climate change is low, unfortunately, due to a weak economy, a complete reliance on oil, and an inability to modernize water infrastructure. Decisions to restructure the agriculture sector will be politically unpopular. What's more, many in the Iraqi government remain unconvinced that the Marshes matter, despite not only their ecological and cultural heritage, but also their role as an important ecosystem for plants and animals, their ability to filter and treat organic wastes, their importance to water supply and livelihood security for over one million Iraqis, and their role in moderating the local climate. The Marshes need to be viewed as part of the solution rather than merely a depository of wastewater.

During the wet period from November to May 2021, rainfall in southwest Turkey—home of the GAP project—was 30 percent below normal. The result was lower harvests and higher food prices. Power production was also being affected, with the country's two largest hydroelectric power plants on the Euphrates River producing at half capacity due to lower water levels in the reservoirs. The lower water levels resulted in Turkey reducing the amount of water flowing into Syria from the Euphrates to 40 percent of the amount guaranteed by the Turkey, Iraq, and Syria Water Accord of 1987. But the damage inflicted on Iraq was much greater than in Turkey. The depth of the Euphrates River at Chibayish was a half meter below normal in early June 2021 and falling at the rate of one centimeter per day. Umm Hussein and others whose lives were dependent on the Marshes didn't know or care what the flow of the Euphrates was on the border between Turkey and Syria. They only knew that water levels were declining, water quality was poor, fish were dying, and dry, cracked ground was appearing where there should have been marsh. In addition, plant growth was inhibited, milk production from the buffalo declined, and buffalo herders had to travel further afield to find fodder for their animals. Some, like Sayed Ismail, were forced to make multiple trips every day to bring fresh drinking water to their buffalo.

Iraq is now considered the fifth most vulnerable country in the world to the negative effects of climate change.[2] Throughout the Middle East, countries are experiencing higher than average temperatures, insufficient precipitation even in the wettest years, increasing frequency of long-term drought, and sand and dust storms. The impacts have been particularly acute for the Marshes and the people living in and around the wetlands. Their problems are worsened by upstream water withdrawals in Turkey, Iran, and within Iraq, water pollution, and inefficient water use. Despite continued warnings from scientists, the United Nations, and other international bodies about the potential impacts of climate change, no one expected the extreme conditions that occurred in 2021–2022—not just in Iraq but throughout much of the northern hemisphere.

Europe experienced its worst drought in 500 years in 2022, resulting in a 20-percent drop in hydroelectric power. Corn, rice, and soybean production were at least 15 percent lower than normal, loss of land from wildfires was the highest ever recorded, and 20,000 excess deaths were attributed to heatwaves. The western United States evidenced the most extreme drought conditions in 1,200 years. Sections of the Yangtze River in China reached their lowest levels since at least 1865. The Eastern Mediterranean and the Middle East fared no better: the year brought the worst drought in 900 years to these regions. The available water in dams and reservoirs inside Iraq decreased from 50 billion cubic meters in March 2020 to less than 10 billion cubic meters in October 2022.

The impacts on the Marshes can be clearly seen from space (figure 31) as well as on the ground: increased desertification, lack of water flow into Iraq from Turkey and Iran, and an almost completely dried out Hawizeh Marsh. Umm al-Ni'aj, once the largest and deepest lake in the Marshes, is nowhere to be seen in figure 31. The deepest section of the lake, known as Abu 'Ethbeh, has been reduced to a few small streams and puddles. However, climate change isn't the only culprit in drying up Hawizeh. The Iranian dike along the border between Iran and Iraq prevents any water from reaching the Iraqi side of Hawizeh. Even during the 1990s when most of the Marshes were drained, Hawizeh survived because of water flow from Iranian streams—but not anymore. Iran claims that its section of Hawizeh (Hor al-Azim) is also drying out, but figure 31 illustrates that this is not the case.

Communities within and adjacent to the Marshes are bearing the brunt of the impact. The village of Muwailih, on the outskirts of Hawizeh Marsh, has been reduced to a few dozen houses of mud and cement surrounded by nothing but dry, cracked land. Boats are strewn about, with no water in sight, and any remaining buffalo appear emaciated. Drinking water for humans and animals, that was once accessed from the nearby al-Mushrah River, must be trucked in from over ten kilometers away. Milk production has decreased, and most residents have left the village. In the distance, one can see flames from gas flaring at the Halfaya oil field—one of the largest in Iraq. The oil is transported via pipeline to Basra and then exported; however, none of the revenues find their way back to villages like Muwailih. One of the residents, Hajj Rikan al-Lami, prays to God for rain but to no avail. He rails at the corruption in government. Hamza, a friend,

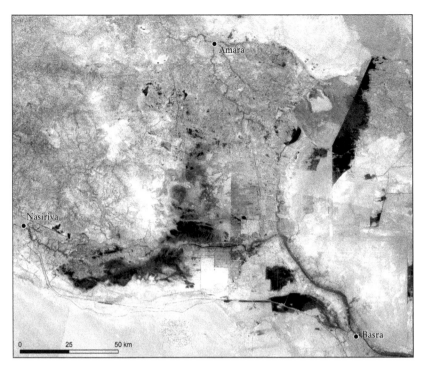

Figure 31. Satellite image of Marsh extent, 2022 (Landsat images courtesy of the U.S. Geological Survey).

tells him to be patient, for the Prophet Ayyub had to wait eighteen years when God afflicted him with an illness.[3] Hajj Rikan only laughs. He tells Hamza that Ayyub was a prophet near the end of his life; he had heaven awaiting him with all that is delicious and good, while their plight is tied to the corrupt politicians. The corrupt ones are destined for the fires of hell.

The changes in Hawizeh Marsh are striking. The wetlands were once teeming with fish life and migratory birds making their way from northern Europe to Africa, but now most species of fish and birds are gone. Many of the villages have been abandoned. There have been public protests by local sheikhs to force the Ministry of Water to open the gates of Nazim al-Kahla', a small river that runs from the Tigris into the Marsh and provides the promised allocation of water to Hawizeh—but with little effect. The people have no choice but to relocate away from the Marsh to survive.

The impacts of the severe drought have not been confined to Hawizeh, however. The Central Marsh is now the driest it has been since 2003, before the Marshes were reflooded. In 2019, the area of wetlands north of the city extended 5,440 km², but by November 2022, they had receded to less than 440 km². Not only has the landscape changed, but buffalo breeders have been forced out of the Marshes, there are few fishing areas remaining, crop production on adjacent lands has decreased, and the local economy is in woeful shape. At least 1,200 families of buffalo breeders and farmers have been forced to leave the Marshes. Some of these families were from the village of Hassjah, a few kilometers north of Chibayish. Umm Hussein moved there in 2015, after being forced to leave her home in al-Ruta, in Hawizeh Marsh. Jassim went to see her in 2022 to ask her how she was coping with the drought. However, Umm Hussein and her family, along with others originally from al-Ruta, had once again become internally displaced. No one could tell Jassim where she had resettled.

The three buffalo breeders were sitting with Jassim in a hastily built mud-and-reed house northeast of Karbala, not far from the Euphrates River in central Iraq (figure 32). The hut lay adjacent to a saline canal and was surrounded by small clumps of brown grass. All three men gazed off into the distance, toward Najaf, where the graves of their fathers lay in the Wadi al-Salam cemetery, reflecting on how their lives had changed in the past half-century. They moved to the region with their families in the spring of 2022, after being forced to sell most of their buffalo when the Marshes once again began to dry out. It was the only way to feed their families. Their few remaining buffalo appeared scraggly and undernourished.

Miri Miaed Medekhel was born in 1968 in Hawizeh Marsh, not far from the Iraq–Iran border. He was only twelve years old when the war with Iran began and his family was forced to migrate to Hammar Marsh. As the war was ending, they moved to the Surah region in the Central Marsh. Three years later, Saddam Hussein's regime drained the Marshes, and Miri Miaed Medekhel and his family—like Najmah, Gumar, and thousands of others— were forced to move away from the Marshes and find work as agricultural laborers near Baghdad. When the Marshes were reflooded, he returned to the Eastern Hammar Marsh. Poor water quality then prompted another move, this time back to the Central Marsh. Now they lived near Karbala, where problems of poor quality in the drainage canals, land tenure issues, and alienation continue to beset the family. Despite these hardships, he still retained hope that the Marshes would return.

Figure 32.
Jassim and the
three buffalo
breeders, 2022
(courtesy of
Rasoul Yahya).

"Without hope, we should not live," he told Jassim.

Buffalo breeders in the Marshes had always been semi-nomadic, moving from one region to another depending on water availability and quality. Being forced to move for political reasons or due to extreme drought added to their hardship. Abu Rassol was born in 1967 in the West Hammar Marsh. His family had been buffalo breeders for generations. Abu Rassol and his seven family members were also forced to migrate when the Marshes were drained in the early 1990s. He returned to the West Hammar Marsh in 2004, but like Miri Miaed Medekhel, Abu Rassol was forced to sell his buffalo in early 2022 and move his family near Karbala.

"Do you think about living in the city?" Jassim asked him. "Wouldn't life be much easier for your family?"

But Abu Rassol was adamant. "No, I cannot live without the buffalo, and I cannot abandon the Marshes."

"But you deserted the Marshes," Jassim said.

"I was forced to leave. God willing, I will return soon."

Abu Jabbar was the third member of the group. He grew up in West Hammar Marsh and became both a buffalo breeder and a fisherman. Abu Jabbar's family was a large one, consisting of thirty members. When the Marshes were drained, he moved to Samawah—a city on the Euphrates River, mid-way between Baghdad and Basra—where he was able to earn a living fishing. Abu Jabbar returned to Hammar Marsh in 2003, only to experience severe drought in 2009, 2015, 2018, and 2021–2022. Once again, he was forced to leave Hammar Marsh. Like his two new friends, he did not despair of the Marshes never returning to their former glory but retained hope of moving his family back to Hammar in the future. He told Jassim that while they were all suffering, they would not be defeated.

No one can predict whether there will be another 700-year drought, but what scientists do know is that there is a much higher probability of drought in any given year than in the past. And these droughts are likely to last longer than before. If this trend continues, as climate models project it will, there simply won't be enough water for the Marshes. The survival of the wetlands has always been subject to the vicissitudes of climate fluctuations; now it also depends on political decisions regarding irrigated agriculture and hydropower in the entire region, as well as Turkey's largesse in meeting its commitments to maintain water flow into Iraq and Syria. It makes the problem of ensuring there is adequate water for the Marshes almost untenable.

Iraq's response to the most recent drought was to release more water from reservoirs behind the Haditha Dam on the Euphrates and the Mosul Dam on the Tigris. This provided at least a brief reprieve from what could have been a catastrophic situation; water flow below the dams is now 50 to 60 percent of normal flow rather than 40 percent or less. However, it is a short-term solution at best. The reservoirs behind the dams are being depleted, and this will have long-term implications if the drought continues—at some point, the reservoirs need to be recharged. The problem for the Marshes is that almost all the additional water being released is destined for agriculture, with precious little left to resupply the wetlands.

Agriculture is an important economic sector in Iraq. While it contributes only 3 percent to Iraq's Gross Domestic Product (GDP), it presently employs just over 16 percent of the population, down from roughly 25 percent in 1991.[4] The economic situation is much different in the southern part of the country, where 30 percent of the people live below the poverty line, and agriculture accounts for over a third of total employment. Agriculture in Iraq requires irrigation, and seventy to 80 percent of all water used in Iraq goes to this one sector. Oil may be king, but agriculture, particularly in terms of water demand and employment, is not far behind. It is a sector that politicians pay attention to.

The main crops in the south are grains—most notably wheat, rice, and barley. Rice is a water-intensive crop, and yields are dependent on how much water is flowing down the Euphrates to Najaf and Qadisiya, the two main rice-producing regions in the country. The government of Iraq is aware of the high amounts of water needed for rice growing and stipulates the amount of land available for rice production on an annual basis, based on the expected amount of water available. In drier years, even with lowered quotas, most of the water reaching southern Iraq is siphoned off for rice production, and very little goes into the Marshes. Despite these measures, rice production in Iraq is highly variable. For example, production was 347,000 tons in 2019 while in the drought year of 2022, it was only 20,000 tons.[5] From the farmers' perspective, *any* water allocated to the Marshes is water that is unavailable for producing rice.

The focus on rice production is perplexing in many respects. At present, 94 percent of rice consumed in Iraq is imported. When Jassim goes to the market to purchase rice, it is typically from Ceylon, India, or the Philippines; rarely does he see Iraqi rice. Indeed, it is difficult to imagine rice being produced in such a water-poor region—it makes no economic or agricultural sense. However, rice producers in southern Iraq wield considerable political power and provide a base of support for some of the Islamist parties. With so many Iraqis in the south involved in agriculture, conflicts over water are not uncommon.

The irrigation system is also a problem in terms of water use. Agricultural areas adjacent to the Marshes are crisscrossed by a stunning array of canals, and these open channels do nothing to limit the evaporation of valuable water. Redesigning the irrigation system is expensive and restructuring the agriculture sector to focus on less water-intensive crops is politically unpalatable. However, there may be no choice—at least not if the Marshes are to survive.

In many areas of the world, the excessive heat and drought linked to climate change are having other economic, social, and ecological impacts. There has been increased heat-wave mortality among humans and other species, damage to crops, and higher demand for electrical cooling which places greater demand on the electrical supply system. There has also been an increase in wildfires which, in turn, have economic, social, and ecological consequences. The wildfires in Canada, the U.S., and southern Europe in recent years have burnt down vast forests, destroyed towns, ruined crops, and killed humans, plants, and animals.

The Marshes are not immune from fire. The Ma'dan have reported fires from Eshan Hafeez for centuries, even though no one dares explore the cause or the source. More common are the fires that are set by residents in the winter months to remove dry strands of reeds and stimulate the growth of young shoots for the buffalo when spring floods occur. The problem is that during drought years the fires can quickly spread, endangering humans and animals. Moreover, the extreme drought of 2018 evidenced a different problem, as fires arose in the summer as well, decimating a large section of Hawizeh Marsh just south of Umm al-Ni'aj and threatening key biodiversity areas. Smoke from the fires affected many villages in and around the Marsh, and northwesterly winds blew heavy smoke into neighboring towns in Iran.

The Marshes of southern Iraq are under assault. Fires, drought, dams, purposeful draining, oil development, and pollution have drastically altered a once unique and productive ecosystem. Work by CRIMW and the Ministry of Water, along with Iraq's involvement in various international conventions, help to ensure the preservation of specific areas of the Marshes in the short term. Nevertheless, regardless of international conventions and the best intentions of government agencies, the Marshes will only survive if there is ample water. In that regard, the prognosis does not look good.

"Would you like to buy some antiquities?" the voice at the other end of the phone asked. The man was telling Jassim about two antique pots that he was willing to sell.

"Where did you acquire these pieces?" Jassim responded.

"A fisherman I know found them on an island in the Central Marsh. He asked me to sell them for him."

Jassim knew that many of the important archeological sites in and around the Marshes had been pillaged by those hoping to sell pieces to international buyers. The General Authority of Iraqi Antiquities had limited resources to excavate the many sites in the region or to pay for security to prevent theft. The resulting concern was that antiquities were being taken out of the country. Working with archeologists such as Dr. Moon and Dr. D'Agostino, as well as with his good friend Abdul Amir al-Hamdani (who was studying for his doctorate in archeology at New York University), Jassim was aware of the antiquities' importance, and he understood the risk of looters to historical sites. He decided to purchase the pieces on offer and donate them to the Nasiriya Museum, which had just reopened in 2015 after being closed for twenty-four years.

One consequence of the violence that has become commonplace in Iraq in the past three decades has been the damage to museums. The important collections at the Nasiriya Museum had to be transferred to Baghdad for safekeeping at the start of the Gulf War, when coalition forces drove the Iraqi army out of Kuwait and the subsequent Shi'a Uprising caused civil unrest in the south. Despite these efforts, there was looting of some important artifacts, including a small clay tablet over 3,500 years old. Written on the tablet is a portion of the *Epic of Gilgamesh*. The tablet, which had been sold and imported into the U.S., was recently returned to Iraq. It is one of over 17,000 pieces that were returned from the U.S. after the Gulf War looting.

The Iraq Museum in Baghdad was less fortunate. It housed one of the most important collections of Sumerian, Assyrian, and Babylonian artifacts in the world, and antiquities experts from many countries pleaded with the U.S. and Britain to ensure the security of the museum during the 2003 invasion. Their pleas came to naught. Fifteen thousand objects were lost in the looting and chaos that followed the fall of the Iraqi government. This could easily have been avoided since there were U.S. and coalition troops stationed a few blocks from the museum. And while some of the artifacts were removed for safekeeping, most of the larger statues remained, and these were easy targets for thieves. Over half of the stolen artifacts have never been returned.

The Mosul Museum, second largest in the country, suffered the greatest devastation of all when ISIS occupied the city in 2015 and destroyed many of the museum's contents. Most items were burned or smashed, but a few of the most valuable pieces were removed and put on sale to raise funds for ISIS. Mosul was recaptured in 2017, and the World Monuments Fund is working with the Iraqi government and donor groups to repair the damage and reconstruct pieces. It was an enormous loss to the region's history and culture.

The looting and destruction of ancient artifacts in Iraq is not confined to museums. Ancient sites throughout the country are being looted for their treasures, many of which are sold on the international market. Jassim was aware of these issues when he received the phone call offering antiquities for sale. The publicity surrounding his advocacy for the Marshes and his international connections—not to mention the photo in which he was mistakenly identified as an American agent—made him an obvious target for vendors of artifacts. The person who contacted Jassim about the sale identified himself as a local government official from Basra. Jassim was skeptical, but decided to research the pieces anyway, in hopes of donating them to the museum. The Nasiriya Museum reopened after most of the items kept in the Iraq Museum in Baghdad for storage were returned. These artifacts were supplemented by over 200 Babylonian-era pieces from Tell Khaybar, a site near Nasiriya. A group of archeologists from the University of Manchester in the UK had excavated the site and restored the artifacts before turning them over to the museum.

Jassim asked the caller to send him photos of the two pieces for sale. He then sent the pictures to his three archeologist friends, asking them to verify their authenticity, importance, and symbolism. All had the same response: the two pots, which were each thirty centimeters in diameter, were covered in Aramaic writing and were called "devil's pots," typically buried in the corner of a house and used to drive away evil spirits. While they were important artifacts, their monetary value was not great—assuming they were authentic.

Jassim immediately called the Basra official and offered to buy them.

"How much do they cost?" Jassim asked, trying not to sound too excited.

"Three million dinars [$2,500]," was the response.

Jassim was not surprised by the request, but pretended the artifacts were not worth the asking price. "I will not pay more than 500,000 dinars [$420]."

Jassim held firm, and the seller finally agreed. The following day, Jassim went to Basra, unsure of what he would find or who the seller would be. He was accompanied by a young driver who could provide security, if needed.

The seller, however, turned out to be very amicable. He prepared breakfast for his two visitors and then handed them the pieces, which Jassim wrapped in blankets for the trip home. When he returned to Chibayish, he called Qais Hussein, the undersecretary at the Ministry of Culture, Tourism, and Antiquities. He told Qais Hussein of his purchase and informed him that he would like to hand over the pieces to the ministry. The two of them later met at the Nasiriya Museum, and now the pieces are part of the museum's permanent collection.

Jassim, however, wasn't quite finished with his research on the two devil pots. In particular, he was interested to learn their provenance. After a few months of investigating and speaking with archeologists, he concluded that the pots were found on Eshan Abu Shadher, an island deep in the Central Marsh. It was a site Gavin Young, an explorer of the Marshes, had visited almost fifty years before. "There was something creepy about it," Young wrote.[6]

The eshan, or island, is 300 meters long, 200 meters wide, and rises 10 meters above the marsh. Like Eshan Hafeez, the site and its treasure were thought to be guarded by tantals. When Young visited in the late 1960s, the Beit Nasrullah tribe inhabited the island and raised buffalo and cattle. The strange and aggressive behavior of the buffalo on the island contributed to Young's unease about the place. No actual treasure was ever found on Abu Shadher, although there had been an early settlement there and residents would uncover artifacts on occasion when they built their reed houses.

The few residents of Eshan Abu Shadher were forced to leave their homes when the Marshes were drained in the early 1990s. The island has since remained uninhabited. The General Authority of Iraqi Antiquities excavated it in 2010 but found little of value and abandoned the site. After Jassim reported on the two pots, he accompanied a television team to the island to film the site, but all they found was a neglected mound surrounded by water. There were no security guards, which once made it easy for thieves to excavate antiquities. During his visit, Jassim found Eshan Abu Shadher somewhat "creepy" as well. Maybe it was the tantals after all.

In his novel *A Banquet for Seaweed*, Haidar Haidar describes the Marshes as a "body of water that lay like an old Sumerian god whose solitude had not been disturbed by mankind for millions of years."[7] By and large, this was true until the middle of the last century, when the Marshes were an expanse of water, artificial islands with reed huts, a few tall and stately palm trees, and a host of fish, birds, and buffalo. Zayir Makki grew up in a reed house on one of these islands in the 1940s. Nearby was the village of al-Khater, close to the Euphrates River and a few kilometers from Chibayish. Standing out from the reed structures was one solitary brick-and-concrete building where the chief of the tribe, Habib al-Khayoun, lived. The village of al-Khater was best known for its two large machines: one for peeling rice, which was the dominant crop in the region, and the other for making ice, an essential product for the fish traders and for keeping goods cool in the hot summer months.

Makki learned to paddle a chileakah when he was only four years old. There were no roads in the district, and once his father enrolled him in primary school, Makki would paddle the boat almost five kilometers each way to class—much like Jassim did. He soon became obsessed with reading stories and books, including the Qur'an. By the time Makki entered middle school, there was a dirt road connecting his village to Chibayish; it made the journey to class much shorter since he could now walk or ride his bicycle. The road attracted food shops, along with a café that entertained both children and adults with one of the few televisions in the area. When he was a bit older, Makki would watch TV in the café, particularly if the boxing legend Muhammad Ali was fighting.

Nevertheless, passage on the dirt road linking al-Khater with Chibayish could be difficult, even in a dry region such as southern Iraq. When it rained, the road was no longer a road, but rather a mud pit. The night before Makki was scheduled to take his annual English examination in middle school, a heavy rain transformed the road into deep mud. Cycling was not possible, so Makki had little choice but to walk. His shoes kept getting stuck in the mud, so he decided to carry them in his arms, along with his schoolbooks, hoping to reach the examination hall before the doors closed. The school was quite strict: any student not arriving at the allotted time would not be allowed to take the test, resulting in failure.

With his legs covered in mud, Makki arrived thirty minutes late. He pounded on the door to the school.

"I am Makki," he yelled, hoping someone inside would respond.

Abbas, the school janitor, heard Makki but told him he could not enter, by order of the school administrator. Makki pleaded with the janitor to inform the director of the school that he was outside. A few minutes later, Abbas opened the door, and the boy nervously entered the hall, tracking mud. Before he reached his desk, the English teacher stopped him. He told the boy that he had no place in the school because the rules do not allow late students to sit for exams.

Makki felt crushed. He complained that the mud delayed his journey, but to no avail. The teacher was adamant that he should not be admitted. Before Makki could argue any further, the director of the school approached, asking him to wash the mud from his feet and sit wherever he wanted. He told Makki that he was worthy of a test because he was one of the best students in the class. Sure enough, Makki passed the exam with high marks, and to this day he reflects on how one's path in life is often determined by a single individual—for better or worse.

Once Makki reached the age of fifteen, he decided it was time to help support the family. He opened a small shop on his island to sell foodstuffs and simple household goods, such as sugar, cigarettes, lentils, and razors—anything people needed. He soon replaced this with a barber shop that attracted residents of neighboring villages, followed by a carpentry shop that catered to people with limited incomes. Makki married when he was twenty-nine years old, ten years later than most of his peers. The first of his children was Salam, who was born in 1968, the fateful year when the Ba'th party seized power in Iraq. Although he was unable to complete his studies, Makki dreamt that his son would eventually go to university, and possibly pursue graduate studies as well.

Like his father, Salam grew up swimming in the shallow canals adjacent to his island home. When he was old enough, he would cross the Euphrates River and enter the Hammar Marsh, where he learned how to paddle a chileakah and the slightly larger mash-huf. Once he completed middle school, his father encouraged him to complete his studies at the Industrial School in Nasiriya, a technical school 100 kilometers from home.

When Salam moved to Nasiriya, he was assigned to federal government housing adjacent to the Euphrates River. Salam had always enjoyed living on the water; it was an important part of his life. Much to his surprise, living next to him was a group of people who not only enjoyed being near the water but worshipped it as well. They were Mandaeans, sometimes referred to as Sabean Mandaeans, and their culture and rituals were immersed in water—figuratively and literally. They believe that flowing, fresh water is the manifestation of all that is good on Earth. It was a revelation to Salam, who realized their belief in the importance of water was close to his own. He would often study in the early morning hours in a stand of palm trees on the bank of the Euphrates, not far from al-Mandada, the temple of the Sabean Mandaeans in Nasiriya. During late March, the Mandaeans would hold the five-day Feast of Parwanaya to celebrate the creation of the world. Salam became fascinated with the role water played in the group's rituals.

The Mandaeans are a monotheistic, gnostic religious group who originated in the Jordan Valley over 2,000 years ago and quickly spread into southern Mesopotamia. They established settlements on the banks of the Tigris and Euphrates Rivers in Iraq and the Karun River in Iran. Adherents follow the teachings of the Prophet Yahya ibn Zakariyya, who is known to Christians as John the Baptist. Baptism is the principal ceremony for Mandaeans because they believe it brings the soul closer to salvation. The practice is performed weekly, and more often during the Feast of Parwanaya. During their ceremonies, the faithful are clothed in white cotton robes, or *rasta*, along with a white cotton turban—both signifying light and brightness in honor of the Prophet Yahya. White, hand-woven belts of wool, called *hymana*, are tied around their waists.

During the Feast of Parwanaya, Salam observed a Mandaean priest walk each member of the congregation into the river and submerge their body three times. The priest then offered three handfuls of water from the river for each member to drink. Salam was moved by the rituals and although he could never join the faith—since conversion is not allowed in the Mandaean religion—the ceremony made Salam feel closer to the Marshes; he knew this was where his soul belonged.

The population of Mandaeans in Iraq and Iran in the 1970s was estimated to be more than 60,000.[8] The Iran–Iraq War, local conflicts, and persecution against religious minorities after 2003 reduced their numbers by over 90 percent. Many became refugees and emigrated to Europe, North America, and Australia.

Salam graduated from the technical school with distinction, which automatically granted him acceptance into the University of Technology in Baghdad, where Jassim studied a few years earlier. He had never been to Baghdad but felt comfortable there. The cement buildings, crowded shops, cafés, theatres, and red British and Hungarian buses were all new and fascinating. He became friends with another student from Chibayish, Hakim Bader al-Asadi, a relative of Jassim's who was also studying electrical engineering. Hakim was more senior and acted as a mentor to Salam, providing him with some needed supplies and showing him around the city.

During Salam's time at university, the country was in chaos. The Iran–Iraq War was just ending, after claiming hundreds of thousands of lives and leaving many orphans and widows. Soon the country was at war again—this time when it invaded Kuwait and was pushed out by the U.S. and coalition forces. On March 5, 1991, the Shi'a Uprising began. Although Salam had never engaged in political or ideological discussions or movements, he felt a need to support his family and friends. Just prior to starting his final year at university, he borrowed a Kalashnikov rifle from a neighbor in Chibayish and took to the streets to support the uprising.

The revolution lasted only a few weeks, and prior to the Iraqi army storming Chibayish, Salam—like Tahseen's family, Adel al-Maajidy, and thousands of others—packed some food and escaped in a small boat through the Hammar Marsh. After three days, he reached the southern bank of the marsh and the highway linking Nasiriya and Basra. From there, he bid farewell to the Marshes and walked for two days before encountering coalition forces and claiming refugee status. Like those before him, he was sent to the Rafha refugee camp in Saudi Arabia.

Salam was more fortunate than many of his friends from Chibayish. His engineering training made him a desirable person for resettlement, and after eighteen months in Rafha, he was offered the opportunity to emigrate to Sweden. Unfortunately, Sweden did not recognize any of his university degrees, and he was therefore forced to repeat courses before applying to a Swedish university. In the end, he not only completed his university degree with distinction, but soon had both an MSc and Ph.D. in Applied Physics Engineering, as well as multiple patents, including one for identifying and decommissioning landmines. But still, he longed for the Marshes.

In 2011, two decades after leaving Iraq, Salam returned with a plan to survey and decommission mines that had been planted around Hawizeh Marsh during the Iran–Iraq War. The mines continued to be a problem for local villagers living in and around the marsh, as they had for Umm Hussein and the residents of al-Ruta. Salam flew to Baghdad and was granted a meeting with the minister of education, who was also a senior member of al-Da'wa, the party in power in Iraq at the time. The minister informed Salam that the government wasn't interested. And although profoundly disappointed, Salam continued his journey to the Marshes, meeting family and friends. He then returned to Sweden to become a director of research and a main advisor to the Swedish minister of defense.

Salam's story is like those of many Iraqi refugees who were forced to escape the country due to political or religious persecution. Quite often, the alternative was jail, torture, and execution. Some refugees, like the Mandaeans, continue their religious practices in countries where religious freedom is a constitutional right. Mandaeans now live together in small neighborhoods in places like Worcester, Massachusetts in the U.S. and the western suburbs of Sydney, Australia. Other Iraqi refugees are fully integrated into the professional and cultural lives of their adopted countries. For those from the Marshes, it is difficult to imagine two more dissimilar environments than the vast wetland in which they grew up, and cities such as Detroit, Stockholm, or Sydney. For them, the Marshes will always be in their blood and soul.

Salam had a dream to return to Iraq and apply his knowledge and experience to improving the lives of those living in the Marshes. Unfortunately, the lack of interest in his project to help remove mines around Hawizeh Marsh is indicative of a broader problem with respect to the preservation of the Marshes: there exists a lack of political will at the highest levels of the Iraqi government to ensure the survival of this unique ecosystem and culture. And while there are professionals like Jassim al-Asadi, Samira Abed al-Shibeb, Dr. Azzam Alwash, and Dr. Hassan al-Janabi advocating for preserving the Marshes, and groups such as Nature Iraq and the Eden in Iraq Wastewater Garden Project lobbying for more resources directed to the Marshes, there are also those with political influence who use their wide reach on social media to declare that allocating water to the Marshes is a waste of valuable resource.

There are reasons why the Marshes have been internationally recognized as worth preserving. Wetland ecosystems are essential in storing and treating water, fostering biodiversity, and moderating climate change, while water reflects solar radiation and plants absorb carbon dioxide. The result is a regional cooling effect. Dry land absorbs more solar radiation and makes a region hotter, and a lack of water can also cause dust storms—as happened in the mid-1990s—with impacts on the health of both humans and animals.

Wetlands are natural filtering systems for agricultural, industrial, and human waste. The Marshes have been effectively treating waste for thousands of years. It is only when waste loads became too large that problems

Figure 33. Traditional dwelling in the Marshes: architecture with an eco-friendly foot-print (courtesy of Mootaz Sami).

arose. Jassim worked for many years to help solve this problem by assisting with the Main Outfall Drain. During the drought of 2009, Jassim partnered with Nature Iraq to direct a portion of the wastewater in the M.O.D. into the West Hammar Marsh to help replenish the wetlands. It was a necessary—albeit risky—experiment, since without the water from the M.O.D., there would not be a marsh. Today, West Hammar is a viable and productive marsh. While the water quality may not meet the standards necessary for human consumption, it still supports a myriad of plant and animal species. Fishing has returned to West Hammar as well—all due to the natural-treatment function of the wetlands.

In addition, wetlands are contributors to socioeconomic development. The Marshes continue to support extensive numbers of fish, migratory and resident birds, amphibians, and plants. When the Marshes were drained, the Ma'dan were not the only group affected; substantial populations of Dhi Qar, Maysan, and Basra depended on the Marshes for sustenance. The number of people living in Chibayish declined by 90 percent in five years as up to half a million people were displaced from their homes. In one way or another, these people depended on the Marshes for sustenance. Nevertheless, although the population within the Marshes never returned to previous levels, there is still a thriving local economy: Chibayish is once again a city of over 60,000; fishing has returned, and total catch equals that prior to the draining; and ecotourism has grown steadily. The Marshes are smaller than before, but they still play an important economic role.

Sustained national development is increasingly being recognized by the international community as having three components: economic, social/ cultural, and environmental.[9] The Marshes embody all three of these components, countering the perception that the wetlands are an undeserving recipient of valuable fresh water from the north. In addition, the Marshes play a crucial role in engaging Iraq with the international environmental community. This is no small achievement, as many countries continue to view Iraq with suspicion due to the high level of governmental/private sector corruption and the uncertain security situation. Preserving the Marshes may be an important step towards improving Iraq's image abroad and enhancing revenues from tourism.

In a water-scarce region, every drop is important—for drinking and cooking, agriculture and industry, energy production, and transportation and recreation. Water is also essential to support ecosystem functions which, in turn, support economic activities such as fishing. The problem is that when fresh water is scarce, it increases the likelihood of conflict among these different uses. Much can be accomplished to prevent this. The agricultural sector is a place to start: improving the irrigation infrastructure and phasing out water-intensive crops, such as rice, would provide both financial benefits and water savings. Enhancing the efficiency of water use by industry is another option; for instance, fresh water should not be injected into wells to support oil reservoirs. Moreover, treatment systems need to be improved. The Marshes are not the culprit in the battle for water, but rather should be viewed as part of the solution.

Admittedly, it is no easy task to convince everyone that the Marshes are important to preserve for reasons of culture, biodiversity, or even ecosystem function—especially in the face of poverty, unemployment, and internal conflict and strife. This is why it is important to have more individuals and groups who are committed to preserving the Marshes and whose voices will be heard at the highest levels of government and in international forums. There is a dire need for advocates for a change in the way Iraq, and, indeed, all countries, balance the importance of nature with powerful economic interests.

Examples of such supporters include photographers and photojournalists such as Meridel Rubenstein, Emilienne Malfatto, and Mootaz Sami, whose striking photographs not only provide an insight into the Marshes and the Ma'dan, but inspire people around the world to want to learn more about the region. There are also archeologists such as Franco D'Agostino, Jane Moon, and Abdul Amir al-Hamdani, who believe our knowledge of the past is essential for understanding our place on this earth, and senior civil servants who have been tireless advocates for the Marshes in the face of both government and private development interests. In addition, there are wetland scientists such as Curtis Richardson of Duke University and international civil servants such as Hassan Partow, who work to understand the ecological, economic, political, and social processes impacting the Marshes, as well as people such as Azzam Alwash and Jassim, who see the need for non-governmental voices to inform the public and galvanize international support for the Marshes' preservation.

There is little doubt that the Marshes, and the people living there, have been under assault for at least the last four decades. Or four thousand

Figure 34. Boy in a *mash-huf* (courtesy of Mootaz Sami).

years. They appear very different today from how they did when Jassim was born. Don't most places in the world? What is so special about the Marshes? Nothing. And everything.

> No one at all sees Death,
> no one at all sees the faces of Death
> no one at all hears the voice of Death,
> Death so savage, who hacks men down . . .
> Ever the river has risen and brought us the flood,
> the mayfly floating on the water.
> On the face of the sun its countenance gazes,
> then all of a sudden, there is nothing there.
> *Epic of Gilgamesh*[10]

The Iraqi Marshes formed over 10,000 years ago and have demonstrated remarkable resilience over the millennia. Despite continuous tussles between Marduk, the god of fresh water, and Tiamat, the goddess of salty water, the Marshes persevered. A multi-year drought destroyed the civilization of southern Mesopotamia, but the Marshes remained long after the great city-states moved elsewhere. Even with the extreme shock to the system from the rapid draining of the wetlands by the Iraqi regime in the early 1990s, the Marshes managed to survive, although there continue to be problems with water supply and water quality. Who is to say that 1,000 years from now the Marshes will not look just as they did a thousand years ago? Or when Abraham was born? Will they be saltwater marshes or freshwater marshes? We can only understand what is happening in the short term, what we can observe and measure. The signs, however, are ominous.

In the above story from the *Epic of Gilgamesh*, Utnapishtim, who has been granted eternal life by the gods for saving humanity from the great flood, is explaining to Gilgamesh how he survived death and became immortal. The passage could also be read as one that presages the life and death of the Marshes. The wetlands were created by the great flood, and as the earth warmed, the sea level rose, and water from the Gulf inundated what is now southern Iraq. Soon thereafter, glaciers began to melt, and runoff from the mountains in southern Turkey swelled the Tigris and Euphrates Rivers and slowly flooded the region, pushing out the salt water and creating a vast wetland. Like the glaciers of 18,000 years ago, those that remain are receding at an alarming rate. Sea-level is again rising and moving up the Shatt al-Arab River. Eastern Hammar Marsh is now a salt-water marsh, and its level fluctuates with the tides. Marduk seems to be losing the battle—there is no longer enough fresh water moving from the north to push back the salty water.

Human presence in the Marshes dates back at least 7,000 years. The first city-state, Uruk, was settled in 4500 BCE. From that time onward, the Marshes became more than a unique ecological area; they developed into an exceptional social, economic, and cultural region as well. The existence of fresh water in southern Mesopotamia allowed human civilization to flourish. The wetlands stored and purified the water and acted as an effective wastewater treatment system, filtering organic matter and nutrients. They were home to a wide variety of plants and animals, fostering biodiversity and helping sustain human life.

Since the middle of the last century, we have learned more about this isolated region from explorers such as Wilfred Thesiger and Gavin Young. Providing valuable insights on the Marshes and the Ma'dan—although admittedly from a Western perspective—their depictions were often tinged with sadness for a life that was changing. In his 1964 book, *The Marsh Arabs*, Thesiger wrote, "Soon the Marshes will probably be drained; when this happens, a way of life that has lasted for thousands of years will disappear."[11]

When he visited the Marshes in the early 1970s, Gavin Young sensed the change as well, much as he felt there was something creepy about Eshan Abu Shadher. Oil was already a major driver of the economy, and at night he could see wells burning off natural gas near Ma'dan villages. Agricultural development, always a threat to the Marshes, was increasing. Coupled with this, a political transformation was occurring that advocated economic modernization over preserving culture or the environment.

The 1992 decision to drain the Marshes was based on the Iraqi regime's desire to repress the local population, transform an area that had served to shelter those fleeing political and religious persecution, and facilitate oil and agricultural development. To these ends, the government bombed and razed villages, resettled the local population to towns and cities, and purposely dried up over 90 percent of the wetlands. When UNEP published its report on the Marshes in 2001, it was clear from satellite photos the extent of the desiccation.

The impact on the Ma'dan and other Marsh dwellers was devastating and irreversible. Those who were not killed or forcibly relocated became refugees or internally displaced persons. They eventually settled abroad or in larger towns and cities in Iraq with better health services, schools, and economic opportunities. Few returned. Poor water quality, more frequent drought, and limited access to services made life in the Marshes untenable for most. Indeed, the intention in draining the Marshes was not simply to dry them out for development; it was a systematic effort to eradicate a people and a culture. In this, the draining was eminently successful.

After the fall of the Iraqi regime in 2003, local initiatives helped to partially reflood the Marshes. Dense reed patches returned, although they were less widespread than in the 1950s. Migratory and resident bird populations rose to 1970s levels. An active Marsh culture remained, although it was much reduced in size. Fishing again became viable, even though the dominant species changed, with an exotic species of carp now accounting for most of the catch. Buffalo numbers are much lower than before the draining, although the remaining animals still wander deliberately into the marshes in the morning and return to their islands late in the day. Their languorous movement may not be in harmony with motorboats, but modernity and the past seem to have reached an acceptable coexistence—except for in years of extreme drought.

Depending on the year and the season, the wetlands now vary between 25 and 50 percent of their size in the early 1970s, with half of this being temporary marsh. Most notably, there have been measurable reductions in water quality, biodiversity, and ecological productivity. And while the Marshes have shown remarkable resilience, they continue to be vulnerable to upstream development and climatic change. Severe drought in 2009, 2015, 2018, and 2021–2022 dried out many shallow wetland areas and increased the concentration of pollutants in the remaining water.

International visitors to the Marshes often fly into Basra Airport and spend two hours in an automobile winding through traffic, noise, dust, and road construction—one could be anywhere in the Middle East or the Maghreb. Then quite suddenly the landscape opens to reveal an immense expanse of reeds and grasses, a myriad of canals, and a few shallow lakes. In the foreground, motorboats are moored or passing to-and-fro, garbage clings to the reeds, and large beasts with horns lounge around or wade into the water. If one gazes in the distance, however, past the symbols of modernity such as cars, electrical wires, and shakhtura with their noisy engines, there is a palpable sense of timelessness, history, and great wonder. The ancient mazes of the Marshes are revealed as welcome intruders in the vast desert. It is a landscape that has endured for thousands of years.

"It's not often that you see a place that reminds you of nowhere else on earth," notes Jane Arraf, the well-known *New York Times* journalist working in Baghdad. "The Marshes are like that. They seem to transcend time, not just in the usual way of a glimpse of life as it was decades or centuries ago but it's more as if everything else falls away out on the water with the birds and the sunlight and the sound of the boat."[12]

Despite their impressive resilience, the Marshes are under threat— of this, there is little doubt. The environment Jassim knew as a boy has been ravaged by social, political, and economic upheaval. Reduced volumes of water in the rivers from upstream dams, agricultural runoff, urban wastewater, large-scale oil development, and a political culture that undervalues preservation of the Marshes combine to offer a bleak picture of their future. The reality of climate change, of drought that can cripple both the wetlands and the people living nearby, casts a pall over not just the region, but the entire globe. The wetlands are now fragmented, and the annual pulse of floodwater that once provided needed fresh water and cleansed the Marshes of pollutants is gone. Today, few people live within the Marshes. Technological change has impacted the region, as it has globally. Shakhtura have all but replaced the traditional mash-huf; aquaculture in the rivers and electric shock fishing are more common than net fishing; and sheikhs no longer use the graceful tarada for transportation, but instead use modern trucks and automobiles.

What changed little over the past half century is the level of economic development in and around the Marshes relative to the rest of Iraq. Widespread poverty remains, and education, health, and other services are still lacking in many areas. To access these services, Marsh dwellers are forced to the edge of the wetlands, relocating to cities such as Chibayish or Nasiriya. Living adjacent to, rather than within, the Marshes has resulted in a cultural shift that has affected everything from lifestyle to building design.

When we see through the eyes of those who have lived the experience of the Marshes, however, there is some hope. The Marshes are now a UNESCO World Heritage site, which has brought international attention to the region's uniqueness and provided strong incentive for the federal government to preserve the Marshes. The World Heritage distinction has brought Iraq one step closer to being fully integrated into the international community. The Mesopotamian Marshlands National Park also exists—if only in name. These two initiatives provide the basis for more eco-development projects in and around the Marshes. Moreover, Iraq is a party to the Ramsar Convention on International Wetlands—a designation that provides Iraq another entryway to international environmental discussions and trust building between nations. The direct impact of these policies and agreements on the people of the Marshes may be an open question, but there is little doubt that they set a framework for local projects that can benefit the Marshes and the people dependent on them.

The Eden in Iraq Wastewater Garden Project is one such example. A joint initiative between Ecotechnics UK/U.S. and Nature Iraq, it is a humanitarian water remediation project designed to demonstrate the use of the Marshes for wastewater treatment. The objective is to take municipal wastewater from Chibayish, currently being discharged into an open canal, and route it through a garden of reeds for treatment before it enters the water channel. The wastewater garden integrates engineering, biology, art, and education in a project that is environmentally and socially responsible. In 2020, the project was included in UNESCO's Green Citizens Initiative. Soon thereafter, the Center for the Restoration of Iraqi Marshlands and Wetlands (CRIMW) agreed to support the pilot project, in conjunction with the district of Chibayish. Meridel Rubenstein (whose photography is included in this volume) and Jassim are two of the principal partners in this initiative. Construction was supposed to begin in October 2021; however, implementation has been postponed due to a combination of inadequate financing and a lack of political will on the part of the Ministry of Water Resources.

Jassim is also involved as a member of the design selection committee for the Museum of the Marshes, located near Chibayish. With funding from the Iraqi Ministry of Culture, Tourism, and Antiquities, construction began in 2009 and the first stage was completed soon thereafter. Piles were installed for a platform extending into the Central Marsh to support buildings that will house cultural artifacts and exhibitions on the Marshes' history. Construction problems, however, derailed progress on the museum, and by the time these were resolved, the country was experiencing difficult economic times and funding for the museum was no longer available. Discussions are presently underway to reinstate funding and complete the museum by 2025.

These initiatives are intended to demonstrate the Marshes' environmental, cultural, and historical importance. They evince how wetland ecosystems are essential in treating wastes, absorbing pollutants, and enhancing biodiversity, and that they are living examples of Iraq's rich history and culture. While the region may also benefit from increased tourism, these projects are not intended to replace traditional Marsh activities, such as fishing and buffalo herding, but aim to protect biodiversity and promote sustainable livelihoods for people.

Nevertheless, implementing these initiatives and projects will never restore the Marshes to their extent of the early 1970s, when the first satellite image of the region appeared. That image of the Marshes provided a moment-in-time view of an ecosystem—and even then, the wetlands were constantly changing. There is one major difference, however. Prior to the 1970s, changes in the Marshes occurred over centuries or millennia. Once humans developed the means to significantly alter the landscape, these changes were significantly accelerated. In the span of fifty years, human activity along the Tigris and Euphrates Rivers and in the Marshes has been nothing short of an assault on the wetlands and the people living there; it is in large part responsible for the near demise of the Marshes. Now, human interventions are needed to restore that ecosystem. Without them, the Marshes will surely die.

There is little question that wetland ecosystems can be vital as water storage systems, despite the evaporation that occurs. Treating wastewater, enhancing biodiversity, and moderating regional climate change are other important features of wetlands. In addition, the Marshes offer an important window to the past and a sense of identity for the future. Preserving natural wetlands makes sense environmentally, economically, and culturally. And yet, preservation of the Marshes is not a high policy priority in Iraq. The country is not alone; over 35 percent of valuable wetlands around the world have been lost in the past fifty years. Unless Iraq makes preserving the Marshes both a domestic and a foreign policy priority, it is difficult to envisage a scenario in which the wetlands survive.

Stronger agreements with other riparian states on the Tigris and Euphrates Rivers, particularly Turkey, are crucial. The problem is that past agreements have been weak and often ignored by upstream countries. This situation will only change if future agreements are tied to Iraq's foreign policy objectives, including those on trade, peace and security, infrastructure development, and climate change. The country's short-term economic fortunes may be based almost entirely on oil, but the availability of fresh water is far more important for the long term. Over half the countries in the world produce no oil, but none exist without a natural supply of fresh water.

Recent years have seen a new wave of diplomacy in the Middle East, as countries realize that a less belligerent attitude brings greater economic and social benefits to the region. Discussions are taking place to break down trade barriers and promote more regional collaboration. For all countries, and Iraq in particular, water should be part of these discussions. Advocating regional water solutions may sound overly optimistic, but when it comes to water, there is no other viable option.

Prior to the Marshes being designated a UNESCO World Heritage site in 2016, the Iraqi committee, supported by the government, made a strong ecological and cultural case for preserving the region. That case needs to be made again. One way of signaling this is to ensure that any discussion of oil and agricultural expansion considers the impact on the Marshes in a meaningful way. Only after the Marshes' preservation is deemed a domestic priority and integrated into other sectors' policies will this issue be taken seriously in international negotiations.

There are also other possibilities for reducing water stress in the country besides better international agreements. As in most countries, the agricultural sector in Iraq uses the most water; it is also the sector that could benefit most from efficiency gains. The 2021 UN report on global food systems acknowledges that agriculture contaminates water and soil and affects human health, that the use of pesticides may be sickening people, and that a focus on rice, wheat, and maize results in epidemics of obesity and chronic disease.[13] The UN, along with many other organizations, also conclude that a move away from water-intensive crops such as sugarcane, wheat, and rice can result in significant water savings, as can the adoption of more efficient irrigation systems. These strategies have been successful in other water-stressed countries and are almost a necessity for Iraq, given the expected decrease in water availability from upstream development and climate change.

Slowly and deliberately, humans are causing a change in the climate system that will have significant effects on coastal areas everywhere. Salt water will continue to push inland. It is hard to conceive that the saltwater marshes which flooded southern Mesopotamia 5,000 years ago will once again cover the region, but it is possible. Eastern Hammar Marsh is already a saltwater marsh. And with less fresh water entering the Marshes from the north and the inexorable advance of salt water from the Gulf, what will the Marshes look like in 2050? How do Iraqis *want* the Marshes to look in 2050? The struggle between Marduk and Tiamat in the Marshes is no longer one of mythology—it has become very real, and now requires careful consideration of how best to intervene to preserve at least a portion of the Marshes.

Improved water agreements and more efficient use of water in agriculture might offset the expected losses due to climate change and upstream dams. But what is needed is a complete transformation of the national water strategy in all sectors to ensure that the Marshes survive. The present political climate in Iraq is unlikely to support such a comprehensive approach, but there are steps that can be taken in the meantime to help the Marshes, with little impact on other sectors. As an initial step, deep water areas need to be protected and reinforced, seeing as how these harbor the greatest biodiversity and provide the most water storage. Reeds that grow in deeper water are the tallest and strongest, and such areas can therefore be sustainable for fishing, buffalo herding, and tourism. The government has gone on record stating that their strategy is to protect the deeper, permanent water bodies in the Marshes and to ensure a minimum of 2,800 square kilometers of marsh is preserved.[14] It is a vacuous promise.

During the summer of 2021, the Iraqi government purposely reduced the flow of water into Hawizeh Marsh to two cubic meters per second, far below the promised amount of thirty cubic meters per second that is essential for the ecological integrity of the marsh. The government claimed that the reduction was necessary to protect drinking water, the one exception made to their promise of protecting deeper water bodies. The action was an attempt to send more water to the Shatt al-Arab River and stem the flow of salt water from the Gulf—a fruitless endeavor to limit the impacts of sea level rise.

The clans in and around Hawizeh had seen enough, and they marched in Amara armed with sticks to force the government to open the gates of the Nazim al-Kahla' regulator, allowing water to again flow to Hawizeh Marsh. The protest portends future conflicts over water in the region for agriculture, fishing, the Marshes, access to drinking water, and fighting saltwater intrusion. In the end, the government acceded, and the flow of water was increased back to the guaranteed thirty cubic meters per second. The problem is that much of this water will be taken for agriculture, with only a third reaching Hawizeh Marsh. Protecting deep water areas and maintaining the 2,800 square kilometers of marsh promised by the government will take more than rhetoric. Regulators and soil embankments need to be constructed to protect existing deep-water areas and allow for water diversions into these areas when needed. Such regions would include Abu Zareg Marsh and Hawizeh Marsh, especially Umm al-Ni'aj, the large lake in Hawizeh.

A second issue is the outflow of water from the various marshes, which is crucial to removing pollutants and stagnant water. Hawizeh Marsh has two natural outlets that help flush pollutants: the al-Kasareh and al-Seeweb Rivers. East Hammar Marsh, which is now a brackish water marsh, is linked to the Shatt al-Arab. Similar outlets are lacking in the Central Marsh and West Hammar Marsh. As a result, there is much higher evaporation in the summer months and a sharp rise in the salinity of the water. Without the pulses of water from seasonal flooding, constructing an outlet in these two marshes is essential for their preservation. Water needs to move through the system.

A third challenge is increasing the use of drainage water to supplement natural flows, using phytotechnology to treat wastewater before it enters the Marsh. Much like the M.O.D. is used to supplement the water in West Hammar Marsh, wastewater can help other marshes, provided there is some treatment. If the Eden in Iraq Project is successful, this could be scaled up to treat polluted water and divert it to the Marshes. Thus, although not without cost, these are relatively simple solutions that would enhance the sustainability of the Marshes and are vital to preserving the wetlands.

There is also an institutional issue that needs to be addressed. The Center for the Restoration of the Iraqi Marshlands and Wetlands (CRIMW) is presently within the Ministry of Water Resources (MoWR). While the mandates of these two institutions overlap, CRIMW's is much broader, and extends to ensuring the sustainability of all wetlands. On issues of water quality, CRIMW works closely with the Ministry of Environment (MoE), whereas on social and economic issues there is sometimes close collaboration with the Ministry of Municipalities. The relationship between the MoWR and the MoE has always been a contentious one, and this places CRIMW in a tenuous and compromised position. The solution to this is relatively easy, although politically controversial, since the MoWR is loath to release control of CRIMW. Nevertheless, the center should be independent from the MoWR and placed under the Council of Ministers. Indeed, there are structural and institutional solutions available to preserve the Marshes; the question that remains is whether there is the political will to do so.

CONCLUSION

A bu Abbas was born in 1930. He grew up in the Marshes playing with turtles and frogs and later earned his living as a fisherman. Abu Abbas knows every meter of the Central Marsh, its islands, its water channels, its clans, and its people. He learned the myths of creation, the stories of lovers, and the tales of ancestors. When he was seventy-five years old, he worked with Jassim and Nature Iraq on a study to help restore the northern Abu Zareg Marsh, forty kilometers southeast of Nasiriya. The Marsh was reflooded in 2003 after being completely dry for eleven years. Abu Abbas saw the water reentering the Marsh, the biodiversity improving, the former villagers returning, and the economic life of the region revitalized.

Jassim went to visit Abu Abbas shortly after his ninety-second birthday. He was weak and in bed, and Jassim asked after his health.

"Praise be to God," he responded. "He wills when He keeps us, and He wills when He decides to pick us like fruit."

"God guard you," Jassim responded. "Your beautiful footprint is everywhere."

Abu Abbas was clearly in much pain and said, "Nothing is eternal but God. But Jassim, what is the status of the Marshes?"

"Things are miserable," Jassim replied without hesitation. "It is nothing like three years ago when the water was plentiful and the reeds abundant."

Abu Abbas then asked about the buffalo breeders in the Central Marsh.

"Most of them have migrated to other places," answered Jassim. Knowing the pain Abu Abbas was feeling, he added, "May the situation improve this winter."

Abu Abbas spoke over him. "The Marshes will not die. They may dry up for one, two, or three years, but they will definitely return as long as there are people who defend them, and I know there are people who defend them."

Jassim admitted that few thought the Marshes would return after the previous regime had dried them up, but they came back thanks to those who believed in them.

"Do not be afraid for the Marshes," Abu Abbas said. "They will survive, even if the water is salty."

Indeed, there is always hope. If there is ample water, the ecosystem will return even if, as Abu Abbas noted, the water might be more saline.

Abu Abbas drew his last breath in early January 2023. Surrounded by friends and family members, he sighed in pain and said, "Preserve the Marshes and fiercely defend your existence. Without the Marshes, you are strangers and homeless." Then his heart stopped beating, indifferent to the cries and wailing from those surrounding him.

It has been three decades since the Iraqi regime dried up the Marshes and displaced almost half a million people. Twelve years later, American and British forces invaded Iraq and toppled Saddam Hussein and his government, allowing local citizens to break down the embankments that restricted waterflow to the wetlands. Since that time, Jassim has been a tireless advocate for the Marshes, initially in his role working for the Ministry of Water Resources, and later with Nature Iraq. His work has helped bring international attention to the region, and he is recognized as a leading authority on both the Marshes and the people living there.

Despite Jassim's efforts, however, the present Ministry of Water Resources has little interest in preserving the Marshes. Environmental issues are a low priority relative to oil and agricultural development, and the biological, cultural, and political values of the Marshes go unrecognized. But government inaction on the Marshes is not the only reason Jassim is worried. Upstream dams, particularly in Turkey and Iran, restrict the flow of water to the Marshes. This is a problem particularly during drought years when there is less water in the rivers, with most of it being siphoned off for irrigation. There have been four major droughts since 2009, and the severity of those droughts is greater than in years past. Politics, dam construction, and climate change give Jassim nightmares. Unfortunately, they are all too real.

In front of Jassim's office in Chibayish, there is a canal that extends from the Euphrates River to the Central Marsh—one of five he proposed to the minister of water resources in 2005 to help rehabilitate the wetlands. The three-hundred-meter canal is adorned with over sixty small fishing boats, and Jassim knows every one of the owners and their families. Theirs is an intimate social group with a strong bond to the Marshes. The boat owners tell him that urban and agricultural pollution is destroying the fish habitats, and their livelihoods are no longer sustainable. In 2021, Jassim's proposal to Minister of Water Resources Mahdi Rashid al-Hamdani to utilize phytotechnology to treat the polluted water fell on deaf ears. The minister refused to meet with an international team working on the project, despite support from the previous minister. It mirrors Salam's experience of working to help remove land mines around Hawizeh Marsh—many in government are apathetic when it comes to the Marshes and those living there.

Jassim no longer has answers to the fishers when they ask about the pollution or an increase in water allocations. Until there is government willingness to preserve the Marshes, the prospects for fishing, buffalo herding, and weaving and construction using local reeds seem dim. Even the survival of the Marshes is in question.

As Jassim sat in front of his office in Chibayish, he fondly remembered building a cooperative relationship with the local population after the 2009 drought, the first one since the Marshes were reflooded in 2003. He encouraged the MoWR to build a soil embankment that would raise the level of the Euphrates River and ensure that water would flow to both the Central Marsh and the West Hammar Marsh, and he saw it through to completion. There is now a forest of reeds stretching to the horizon, broken only by the boat channels and small groups of buffalo wading through the water. The songs of the reed harvesters can be heard in the distance, and Jassim recalls his first love in these marshes. His thoughts were interrupted as a slender woman paddled a boat toward him.

"Don't forget us. We need more water," she said.

Jassim looked at his small yellow kayak that lies under a palm tree in the garden of his office building. A German friend gave him the kayak over a decade ago, when Nature Iraq helped organize a protest in the Turkish city of Hasankeyf over the construction of the Ilisu Dam on the Tigris River. Jassim enlisted help from locals in the Marshes and support from

European journalists and international environmental activists in an effort to stop the construction. In Hasankeyf, they camped in a lush orchard on the bank of the Tigris. On the opposite bank were hundreds of empty caves carved into the rock, sitting high above the river. People had lived in the caves for almost 4,000 years, through to the 1970s. The reservoir behind the Ilisu Dam submerged the ancient city in 2020. Contrary to Turkish law, no environmental impact assessment was ever undertaken on the project, and pressure to make Hasankeyf a UNESCO World Heritage site was ignored by the government. Three European donor agencies withdrew from the dam project due to Turkey's failure to meet international standards of protecting nature, culture, and the rights of more than 25,000 people who were eventually displaced by the project. The protestors met with the regional governor, but there was little he could do.

The completion of the Ilisu Dam was a great disappointment for Jassim. He knew the dam would destroy an important cultural and historical site in Turkey and, more importantly, he recognized the impact the dam would have on the Marshes. Ilisu was part of a series of dams designed to produce hydroelectricity and irrigate agricultural land, reducing the amount of water that historically flowed downstream. Despite international pressure and court cases within Turkey designed to halt dam construction, Jassim was under no illusions that the project would be cancelled. But for the sake of the Marshes, he had to try.

Another disappointment for Jassim occurred in 2009, when Iran completed a 64-kilometer embankment along the border with Iraq, stopping the flow of water into Hawizeh Marsh. The purposes of the embankment were to divert water for irrigation within Iran, allow for more oil development, and prevent Iraq from accessing the water. Hawizeh was the one marsh able to survive the ravages of the Iraqi government's policy of draining the wetlands because of the water flowing from Iran. The embankment put an end to this and meant that the only source of water to the marsh would be the Tigris River.

Jassim wrote letters to the press and organized a trip for Iraqi media to the site, which resulted in a movie that attracted a wide audience. Iraq, however, wasn't interested in engaging with Iran over the issue. And although Jassim did participate in a UN-sponsored meeting about the Marsh, the Iranians were unwilling to discuss the common issue of marsh preservation. The only project that came out of the meeting was a joint survey of birds in the marsh, which Jassim proposed in the hope of engaging Iran on at least one issue and furthering trust-building between the countries.

In early 2022, as he gazed at the canal in front of his office, Jassim thought about the best and worst days of his career. What did he do right? What did he regret? The best decision he made was to retire early from the Ministry of Water Resources and work full time with Nature Iraq. This led to two major achievements. The first was his proposal to use water from the Main Outfall Drain to flood large areas of the West Hammar Marsh after the drought of 2009. Although polluted with agricultural runoff, the M.O.D. saved West Hammar. When the water came, so did plants that absorb organic pollutants, and the water quality later improved enough to allow fishing and reed harvesting. Today, the marsh is once again a dynamic ecosystem and supports many economic activities. The second achievement was Jassim's recommendation to build an embankment along the Euphrates to allow more water to flow to the Central Marsh and the West Hammar Marsh. Although smaller than when he was a boy, these two marshes once again have a source of water for replenishment—at least during years of average or above average water levels in the Euphrates River.

Jassim is a technical engineer whose greatest pleasure is being in the Marshes, surrounded by people and nature. Not only does he have an extensive network of social relations in the region, but he is a quintessential educator on, and advocate for, the Marshes. Whether in formal speaking engagements or in leading trips for journalists, diplomats, and professionals interested in knowing more about the region, Jassim is the first point of contact. However, his life is not without disappointments. Setting aside the physical and mental pain he continues to experience from his months of torture and the grieving for good friends who died or left the country during times of conflict with the Ba'th regime, he also has professional regrets. He is most disappointed with the poor implementation of the management plan for the UNESCO World Heritage site and the overall failure of the Ministry of Water Resources to manage drought in the wetlands.

Jassim also regrets unintentionally misleading Marsh dwellers regarding the willingness of the ministry to help manage the Marshes. He recalls how bewildered the buffalo breeders were when the Marshes dried up on four occasions in recent years. They watched perplexed as the quality of water deteriorated, the price of feed rose, and the buffalo produced less milk. Like Umm Hussein, they had little understanding of what was causing the changes. Climate change, geopolitics, and even conflicts at the national level between resource ministries seemed too complex for the simple life of the Ma'dan. Like Jassim's grandmother, they believe that their enemies live somewhere in the wastelands and that the angels will protect them.

Jassim also experienced first-hand how little previous ministers of the MoWR were willing to do for the Marshes. He acknowledges that until there is someone in the post who is either in love with the Marshes or understands their social, environmental, and cultural importance, there is little hope for the future. Nevertheless, there is at least one reason for optimism. Jassim has experienced the role of local initiatives as drivers of change: from the reflooding of the Marshes in 2003, to the recent protests against the government for reducing water flow to Hawizeh, he has seen how grass roots efforts cause disruptions and effectuate transformation. Granted, they don't always work—as he saw first-hand with the Shi'a Uprising and the efforts to stop the building of the Ilisu Dam—but they can have an impact. As long as he is able, Jassim will continue to try.

The Ma'dan believe that God sent tantals down from the heavens to guard the ancient cities of the Marshes and the treasures they contained. Jassim is still wary of them. Tantal Hafeez, who protects Eshan Hafeez, is the most well-known and the most dangerous—even non-believers stay clear of Eshan Hafeez. Tantals both haunt and protect the Marshes. How else can one explain the survival of the Marshes through floods, drought, fires, and the malevolent actions of human beings? Just as the gravedigger experienced in Wadi al-Salam, tantals are now hitting people over the head, in a vain attempt to get us to listen. Destroying an ecosystem that not only is ecologically and culturally unique, but also economically and socially valuable will have far-reaching impacts, from lost biodiversity to dust storms. The tantals will not leave quietly. They may well haunt the souls of those in the Marshes forever.

The Garden of Eden, the scene of the great flood, the birthplace of Abraham, the place where writing and mathematics were invented and civilization began, home of hydraulic engineering, textile manufacturing, the plow, and mass-produced bricks . . . The list goes on. The story of the Marshes and the people living in southern Iraq is one of hardship and wonder, cataclysm and survival. Almost every human being in the world is in some way linked to the Marshes, regardless of religious affiliation,

Figure 35. Jassim (sketch by
Bassim Al-Asmawi, 2022).

nationality, or ethnicity. Fifty-five percent of the world's population iden-
tifies themselves as Christian or Muslim and their beliefs are associated
with this region, while most other religions have creation and flood myths
that link back to the *Enuma Elish* and the *Epic of Gilgamesh*. And while writing, metallurgy, hydraulic engineering, and mathematics later developed in
other parts of the world, they will always be linked to the great city-states
of Ur and Uruk.

It may not matter, however, since climate change could ultimately
destroy what is left of the Marshes. Saddest of all is the potential loss of
the unique culture of the Ma'dan. Already the disruptions and disloca-
tions caused by the draining have devastated the lives and traditions of the
Marsh dwellers. With the loss of culture goes the loss of arts and of stories.
The great Ma'dan storytellers are, for the most part, gone. With them have
gone stories about the Ma'dan, how they experienced history, and how
they understood the natural order of the world.

It is true that the modern history of the Marshes has been a tragic one; however, as with Jassim whose childhood poem ends with "I will not die," the Marshes also have not died. May they survive for another ten thousand years, *inshallah*.

AFTERWORD

Steve Lonergan

It should be no surprise that being an environmental activist in Iraq, a public advocate for water use and water rights, and a champion for ensuring the Marshes receive enough water to survive, carries a high degree of personal risk. On February 1, 2023, Jassim was kidnapped while driving with his cousin Raad from Hilla to Baghdad for a meeting with the president of Iraq. Two cars blocked the road, and five men approached Jassim's car with their guns drawn. They pulled Jassim out of the car, covered his face, gagged him, tied his hands, and dragged him into another vehicle. He was shoved roughly between two seats, and the two cars quickly departed. Raad was left unharmed in the car and was able to see both the kidnappers' faces and the license plate numbers on one of the cars.

Jassim's family had no idea why he was taken or what the conditions might be for his release. The license plate gave the family and senior members of the Bani Asad a crucial piece of evidence regarding who might be responsible. They believed the kidnappers were members of an armed militia from a cabal that has increasingly exerted its influence over social, economic, political, and military activities in the country, particularly in southern and central Iraq and some areas bordering Iraqi Kurdistan. When the U.S. invasion in 2003 left the country in chaos and the Iraqi government in disarray, it presented an opportunity for a few non-state actors, often with funding from outside Iraq, to increase their power. To exert their authority, they engaged in killing, looting, and kidnapping, and their targets were often people with links to the West. Some of these groups are now represented at senior levels of government.[1]

Five days after the kidnapping, having heard nothing from his abductors, the police, or the federal government, Jassim's brother Nazem approached AFP, the global news agency in Iraq, with the story. Within hours of the press announcement, condemnation of the kidnapping appeared on social media, as people speculated on why Jassim was kidnapped, who was responsible, where he might be held, and what the government was doing to address the situation. Prominent leaders such as Hassan Al-Janabi, Iraq's former Minister of Water Resources, and Inger Andersen, executive director of the UN Environment Programme, posted strong statements of support for his release. The German Embassy in Baghdad published a statement condemning the kidnapping and holding the Iraqi government responsible for saving Jassim's life. The Iraqi Observatory for Human Rights cautioned all activists and human rights defenders that his disappearance suggested a possible return to systematic targeting.[2]

The Iraqi government, concerned about the possibility that tensions between Jassim's Bani Asad tribe and the militia responsible for the kidnapping might escalate into violent conflict, appealed for calm and made it clear that they were in control of the situation. As the days went by with no further news, Nazem complained about the lack of communication from the government. Twelve days after the kidnapping, Iraqi Prime Minister Muhammad Shayya' al-Sudani promised that his security forces would rescue Jassim and arrest those responsible. It was a promise that gave little comfort to many of his friends outside the country.

Three days later, Jassim was released in Baghdad to his brother's care. He remained in Baghdad for the night and then was reunited with his family the following day in Chibayish. A few hours after his release, I received a text message from a mutual friend saying he was free. Word spread quickly and, based on the various emails I received and posts on Jassim's Facebook page, I sensed a collective sigh of relief. Many of us had feared the worst.[3]

Jassim called me via a video link late in the evening (Iraqi time) the day after his release. Appearing gaunt and pale and sounding very weak, he inquired whether I had heard the news of his kidnapping. I told him yes, and that little else had crossed my mind for almost two weeks. He told me he would be resting for a few days in Chibayish and then traveling to a hospital in either Dubai or Amman for treatment. Jassim then asked, "Are there any more news about the book?" If there was a positive sign in our conversation, this was it. I reinforced that the book would be published in the coming months and encouraged him to do nothing but take care of himself and get healthy.

Jassim's release was celebrated a few days later at an outdoor cultural event in Nasiriya. Part of the event was to recognize writers, artists, and others who supported the Iraqi environmental movement and the preservation of the Marshes. He was later interviewed separately by Al Sharqiya TV and Almada, a leading newspaper in Iraq. After thanking those who facilitated his release and those friends and supporters who were there to greet him on his return to Chibayish, Jassim admitted he had been subjected to daily torture. It was like reliving the nightmare he endured under the Ba'thist regime when he was jailed for nine months, with daily beatings, electrical shocks, and waterboarding, among other methods of torture.

Despite his ordeal, Jassim remained resolute. "The support I have received . . . makes me stick to the cause for which I was kidnapped, and I will continue my same path of defending the environment [meaning the Marshes] and its inhabitants," he told the audience. Jassim again demanded that neighboring countries ensure Iraq receives its fair share of water from the Tigris and Euphrates rivers and that the Iraqi government, in turn, keep its promises to supply an adequate amount of water to the Marshes to alleviate the suffering of the local population.

After the kidnapping, Jassim spent a few days recuperating in Chibayish before going to Amman, Jordan, for a week of medical treatment. On the invitation of former Iraqi President Barham Salih, he then spent two weeks in Sulaymaniyah in northern Iraq where he participated in discussions on water use and the plight of the Marshes.

On March 12, 2023, at a climate change conference in Basra, Jeanine Hennis-Plasschaert, special representative of the UN Secretary-General for Iraq, made the following statement. Sitting next to her was Iraqi Prime Minister Muhammad Shayya' al-Sudani.

"I am very happy to be here, and if you allow me, I would like to start my speech by paying tribute to Mr. Jassim Al-Asadi. His lifelong commitment to preserving the environment is not only appreciated, but also essential. We need people like Mr. Jassim. They open our eyes. They wake us up. They are critical drivers for change, adaptation, and progress."[4]

It is still unclear why Jassim was kidnapped. He believes it was because of his many contacts with researchers, journalists, and activists in the West, coupled with his environmental activism in Iraq. Following his stay in

Sulaymaniyah, Jassim returned to Chibayish. Surrounded by friends, his wife Suad, and a few other family members, he feels safe from the dangers that almost cost him his life. He still makes the occasional trip to Babylon and Baghdad, but only for short periods. Jassim's health continues to improve, and I hope he might be able to obtain a U.S. or Canadian visa to visit and help launch our book. It would be wonderful to introduce him to a North American audience. In the long term, however, Jassim is adamant that he will remain in Iraq and work for the preservation of the Marshes and the welfare of its people. As he told me, "Jassim without the Marshes is not Jassim."

GLOSSARY

'angar: The youngest reeds (less than six months old) that are eaten by buffalo.

chileakah: A very small, bitumen-covered canoe (smaller than a mash-huf; see below).

eshan: Large islands that rise high above the wetlands.

guffa: Circular, flat-bottomed boats that look much like large baskets covered with leather. They have been used in Iraq (and elsewhere) for over two thousand years to shuttle people and goods back and forth across rivers.

hashish: Another word for 'angar, the youngest reeds eaten by animals.

kalak: A raft made of logs supported by inflated goatskins and used to carry people, animals, and goods downriver.

mash-huf: A bitumen-covered wooden canoe for two persons that is used for general transportation, fishing, and collecting reeds.

masnayat: Fences of reeds that surround buildings to protect them from floods.

mudhif: A large structure built from arched reeds that is used for meetings, community gatherings, and as a guest house.

rawat: The generations of storytellers in the Marshes.

Salwa: The demonic spirit in Ma'dan mythology, similar to the better-known Lilith.

shibab: Large arches of reeds that form the structure of a mudhif.

tantals: Mythical creatures that keep watch over the marshes.

tarada: A large canoe—up to 12 meters long—made of wood and covered with bitumen. The stern and the bow are tapered and curve up.

NOTES

1   Jane Arraf, email response to questions about the Marshes from the author, September 20, 2021.

NOTES TO THE INTRODUCTION
1   Formerly the Arabian Gulf or the Persian Gulf, now simply called "the Gulf."
2   Wilfred Thesiger, *The Marsh Arabs* (London: Longmans, Green, & Co., 1964).
3   Gavin Young, *Return to the Marshes: Life with the Marsh Arabs of Iraq* (London: HarperCollins, 1977).
4   Young, *Return to the Marshes*, 219.
5   Hassan Partow, *The Mesopotamian Marshlands: Demise of an Ecosystem* (Nairobi, Kenya: United Nations Environment Programme, 2001).

NOTES TO CHAPTER ONE
1   al-Jahiz, *al-Bayan wa-l-tabyyin* (Windsor: Hindawi Foundation, 2022), 139, quoted in "al-Hakawati: A Digital Library of Arab and Islamic Culture. Tribes and People: Bani Asad," accessed June 22, 2021, http://al-hakawati.net/en_cultures/CultureDetails/609/Bani-Asad.
2   al-'Awwad is a sub-tribe of the Bani Asad. One of Jassim's uncles, Abu Abbas, was their tribal leader.
3   Jassim al-Asadi, "I Will Not Die" (unpublished, 1969).
4   "The Debate Between Bird and Fish: Translation," The Electronic Text Corpus of Sumerian Literature, University of Oxford, accessed August 1, 2021, https://etcsl.orinst.ox.ac.uk/section5/tr535.htm#:~:text=In%20

those%20ancient%20days%2C%20when,his%20hand%20waters%20
to%20encourage.

5   Leonard W. King, ed., *The Seven Tablets of Creation, or the Babylonian and Assyrian Legends Concerning the Creation of the World and of Mankind* (London: Luzac & Co., 1902), 4.

6   "al-Rawat" in Arabic translates to "the storytellers."

7   The Maghreb is Northwest Africa. One might infer that the "seven earths" refers to the continents.

8   "Gravediggers Claim Ghosts Haunt World's Largest Cemetery in Iraq," Al-Jazeera, accessed July 20, 2021, https://www.aljazeera.com/features/2019/9/10/gravediggers-claim-ghosts-haunt-worlds-largest-cemetery-in-iraq#:~:text="One%20day%2C%20a%20shadow%20sneaked,nicknamed%20Tantal%2C%20Bzebza%20or%20Ghreria.

9   Samir Naqqash, "Tantal," in: *Contemporary Iraqi Fiction: An Anthology*, ed. Shakir Mustafa (Syracuse: Syracuse University Press, 2008), 115–129.

NOTES TO CHAPTER TWO

1   Gibran Kahlil Gibran, "Fear," accessed September 10, 2021, https://yourdailypoem.com/listpoem.jsp?poem_id=3608.

2   Jassim al-Asadi, "Let Me Narrate From Your Cheeks" (unpublished, 1963).

3   Arab nationalism or Pan-Arabism promotes the belief that Arabs should unite to form one state. Since 85–90 per cent of Arabs are Sunni Muslims, other groups, such as Kurds and Shi'a Muslims, are adamantly against Arab nationalism.

4   Safwan is a Qur'anic name that means "rock."

5   Manjeri Subin Sunder Raj, "The Legal Lore of Water Ecology and Scriptures," in *Water and Scriptures: Ancient Roots for Sustainable Development*, ed. K. V. Raju and S. Manasi (Cham: Springer, 2017), 173–217.

6   "The Code of Hammurabi," The Avalon Project, accessed October 5, 2020, https://avalon.law.yale.edu/ancient/hamframe.asp.

7   Dante Caponera, "Ownership and Transfer of Water and Land Tenure in Islam," in *National and International Water Law and Administration: Selected Writings* (Boston: Brill, 2021), 73–80.

NOTES TO CHAPTER THREE

1   "Haidar Haidar's 'Hymns of Death': Translation by Allen Hibbard and Tadween Editor Osama Esber," Tadween Publishing, accessed December

1, 2020, https://tadweenpublishing.com/blogs/news/haidar-haid-ars-hymns-of-death-translation-by-tadween-contributor-osama-esber.

2   Human Rights Watch, *Iraq: State of the Evidence*, 2004, https://www.hrw.org/reports/iraq1104.pdf.

3   Iraq previously broke off relations with Egypt after Anwar Sadat's visit to Israel in 1977.

4   Jerome Donovan, *The Iran-Iraq War: Antecedents and Conflict Escalation* (London: Routledge, 2014), 90.

5   Bruce Riedel, "Foreword," in *Becoming Enemies: U.S.-Iran Relations and the Iran-Iraq War, 1979–1988*, eds. James G. Blight et al. (Lanham: Rowman and Littlefield, 2012), ix.

NOTES TO CHAPTER FOUR

1   Mourid Barghouti, "Midnight," in Midnight and Other Poems, trans. Radwa Ashour (Lancashire: Arc Publications, 2008), 70.

2   "Why Does Allah Love Odd Numbers (Witr)?," al-Hakam, accessed September 25, 2022, https://www.alhakam.org/why-does-allah-love-odd-numbers-witr/. This is a hadith by the Prophet. Since Allah is One (an odd number), it is believed that He loved odd numbers. For example, it took seven days to create the world. Accordingly, odd numbers are seen as special in Islam.

3   Jim Hoagland, "Transcript Shows Muted U.S. Response to Threat by Saddam in Late July," *The Washington Post*, September 13, 1990, https://www.washingtonpost.com/archive/politics/1990/09/13/transcript-shows-muted-us-response-to-threat-by-saddam-in-late-july/790093d9-92f6-4906-87b3-630aab59e9b7/.

4   United Nations, *Report to the Secretary-General on Humanitarian Needs in Kuwait and Iraq in the Immediate Post-Crisis Environment by a Mission to the Area Led by Mr. Martti Ahtisaari, Under-Secretary-General for Administration and Management*, March 20, 1991, 6, https://www.un.org/depts/oip/background/reports/s22366.pdf.

5   Based on a conversation I had with Professor Eran Feitelson, a professor at Hebrew University and resident of Jerusalem, October 10, 1994.

6   "Remarks to the American Association for the Advancement of Science," George H. W. Bush: Presidential Library and Museum, accessed February 1, 2021, https://bush41library.tamu.edu/archives/public-papers/2709.

7　Based on an unpublished video interview conducted with Ihsan Kadhim amml-Asadi on June 23, 2021.

8　Based on unpublished telephone interviews with Adel al-Maajidy and his daughters Asma and Assra, April 13 and 20, 2021.

9　United Nations Digital Library, *UN Security Council Resolution 688, Adopted on 5 April 1991*, April 5, 1991, 34, https://digitallibrary.un.org/record/110659?ln=en.

10　Fred Pearce, "Draining Life from Iraq's Marshes: Saddam Hussein is Using an Old Idea to Force the Marsh Arabs from their Home," *New Scientist*, April 17, 1993, https://www.newscientist.com/article/mg13818691-800-draining-life-from-iraqs-marshes-saddam-hussein-is-using-an-old-idea-to-force-the-marsh-arabs-from-their-home/.

11　Shyam Bhatia, former diplomatic editor of The Observer (UK), was named the 1993 Foreign Journalist of the Year at the British Press Awards for his coverage of the Marsh Arabs.

12　Human Rights Watch, *The Iraqi Government Assault on the Marsh Arabs: A Human Rights Watch Briefing Paper*, January 2003, https://www.hrw.org/legacy/backgrounder/mena/marsharabs1.htm.

13　A sluice gate is a moveable gate that can be open or shut to alter the water flow in a canal.

14　Less than 1,000 parts per million is considered fresh water.

15　Unpublished interview with Najmah, June 14, 2021.

16　The opening seven verses of the Qur'an, which serve as a prayer for guidance, lordship, and the mercy of God.

NOTES TO CHAPTER FIVE

1　Omar F. Al-Sheikhly and Iyad A. Nader, "The Status of Iraq Smooth-Coated *Otter Lutrogale perspicillata maxwelli* Hayman 1956 and Eurasian Otter *Lutra lutra* Linnaeus 1758 in Iraq," *IUCN/SSC Otter Specialist Group Bulletin* 30, no. 1 (2013): 18–30.

2　Canada-Iraq Marshlands Initiative, *Managing for Change: The Present and Future State of the Marshes of Southern Iraq*, 2010, 17.

3　Center for the Restoration of Iraq Marshes and Wetlands, *Study of the Impacts of the Oil Industries on the Marshes of Southern Iraq, Final Report*, 2014, 24.

4　Ernestina Coast, "Demography of the Marsh Arabs," in *The Iraqi Marshlands: A Human and Environmental Study*, eds. Emma Nicholson and Peter Clark (London: Politico's Publishing, 2003), 20.

5   I was working in Nairobi as director of the science division of the
    UNEP at the time and recall very clearly when a staff member burst into
    my office and told me the UN complex in Baghdad had been bombed.
    "But Hassan is OK!" she exclaimed. I had no idea one of my staff—from
    Geneva—was in Baghdad. The bombing in the Canal Hotel, along with
    the bombing of the U.S. Embassy in Nairobi only five years before—
    also by al-Qa'ida—raised the threat level at the UN Offices in Nairobi
    (which were adjacent to the new U.S. Embassy) to very high. We were
    subjected to daily bomb checks before entering the UN complex. - SL
6   Iraqi Const. art. II. Adopted October 15, 2005.
7   Ms. Israa Abd Al Ali Abdul Wahab, Second Deputy of the Basra Provincial
    Council, told me this at a meeting on the Marshes in 2006. A *niqab* is a veil
    that covers a women's face and body with only a small opening for the eyes.
8   UNHCR, *Iraq Situation Response: Update on Revised Activities Under the
    January 2007 Supplemental Appeal*, July 2007, 1, https://www.unhcr.org/
    media/iraq-situation-response-update-revised-activities-under-janu-
    ary-2007-supplementary-appeal.

NOTES TO CHAPTER SIX
1   The Convention on Wetlands is named after Ramsar, Iran, where the
    convention was signed in 1971.
2   Fatimah Naeem Dherib, "Establishing the City of Nasiriyah," *PalArch's
    Journal of Archaeology of Egypt/Egyptology* 17, no. 7, (2020): 6427. Sheikh
    Nasir al-Saadoun was later known as Sheikh Pasha Nasir al-Ashkar. In
    exchange for large tracts of land, he agreed to cooperate with the Otto-
    mans and was soon appointed district governor. The city of Nasiriya is
    named after him.

NOTES TO CHAPTER SEVEN
1   Tadween Publishing, "Haidar Haidar's 'Hymns of Death.'"
2   A weir is a low dam that allows flood water to flow over the top.
3   Alan Taylor, "An Ancient Town Submerged: Hasankeyf Underwater," *The
    Atlantic*, October 1, 2020, https://www.theatlantic.com/photo/2020/10/
    photos-an-ancient-town-submerged-hasankeyf-underwater/616562/.
4   Anonymous, "The American Intelligence Officer for Middle East Affairs,
    Richard Funk," Facebook comment, September 25, 2018. Purple finger
    is a reference to the U.S. controlling the upcoming elections in Iraq.

5   Dr. Ahmed Amin, "The fruitful trees are always stoned," Facebook comment, September 27, 2018.

6   Mudher Muhammad Salih, "al-'Iraqi al-ashqar: niqma am ni'ma?," *Ahewar*, October 4, 2018, https://www.ahewar.org/debat/show.art. asp?aid=613675.

7   "Nature Iraq," Nature Iraq, accessed June 10, 2021, http://www. natureiraq.org/read-how-the-conservation-community.html.

8   Consuming alcohol or any intoxicant is considered *haram* (forbidden) in Islam. The Prophet Muhammad spoke against drinking alcohol, partly because it drives people away from God. A minority of Muslims— including Fadi—continue to consume alcohol, and the Qur'an does not state a specific punishment for the act.

9   "The Atlas of Economic Complexity," Growth Lab, accessed May 15, 2021, https://atlas.cid.harvard.edu.

10 Simon Watkins, "The Real Reason Why ExxonMobil Won't Go Ahead With $53 Billion Iraqi Megaproject," Oil Price, accessed April 4, 2021, https://oilprice.com/Energy/Energy-General/The-Real-Reason-Why-ExxonMobil-Wont-Go-Ahead-With-53-Billion-Iraqi-Megaproject.html.

11 "Iraq Pipeline Watch," Institute for the Analysis of Global Security, accessed April 30, 2021, http://www.iags.org/iraqpipelinewatch.htm.

12 Nature Iraq, *Key Biodiversity Areas of Iraq: Priority Sites for Conservation and Protection*, (California: Tablet House Publishing, 2017).

NOTES TO CHAPTER EIGHT

1   Climate Central, *Global Weirdness: Severe Storms, Deadly Heat Waves, Relentless Drought, Rising Seas, and the Weather of the Future* (New York: Pantheon Books, 2013).

2   United Nations Environment Programme, *Global Environment Outlook – GEO-6: Healthy Planet, Healthy People* (Cambridge: Cambridge University Press, 2019), https://www.cambridge.org/core/books/global-environment-outlook-geo6-healthy-planet-healthy-people/8FE2F-127F310561C679B620F1D2EDBA6.

3   Prophet Ayyub, known as Job in the Bible, was considered the most loyal servant of God and one of the most patient. He was blessed with many riches but later was afflicted with a flesh-eating disease and lost his family and his possessions. Despite these hardships, he never lost his patience and remained faithful to God.

4 "Iran: Employment in Agriculture," The Global Economy, accessed July 15, 2023, https://www.theglobaleconomy.com/Iran/Employment_in_agriculture/#:~:text=Employment%20in%20agriculture%2C%20%25%20of%20total%20employment&text=For%20that%20indicator%2C%20we%20provide,from%202021%20is%2016.25%20percent.

5 "Iraq Rice Area, Yield, and Production," United States Department of Agriculture, accessed July 15, 2023, https://ipad.fas.usda.gov/country-summary/Default.aspx?id=IZ&crop=Rice.

6 Young, *Return to the Marshes*, 40.

7 Tadween Publishing, "Haidar Haidar's 'Hymns of Death.'"

8 Shak Hanish, "The Mandaeans in Iraq," in *Routledge Handbook of Minorities in the Middle East*, ed. Paul S. Rowe (London: Routledge, 2018), 159–169.

9 United Nations Environment Programme, *Global Environment Outlook*.

10 Michael Schmidt, *Gilgamesh: The Life of a Poem*, (New Jersey: Princeton University Press, 2019), 80.

11 Thesiger, *The Marsh Arabs*, 11.

12 Jane Arraf, email response to questions about the Marshes from the author, September 20, 2021.

13 Food and Agriculture Organization, *The State of Food Security and Nutrition in the World: Urbanization, Agrifood Systems Transformation and Healthy Diets Across the Rural–Urban Continuum* (Rome: FAO, 2021).

14 Charlotte Bruneau and Thaier Al-sudani, "'Our Whole Life Depends on Water': Climate Change, Pollution and Dams Threaten Iraq's Marsh Arabs." *Reuters*, 14 October 2021, https://www.reuters.com/world/middle-east/our-whole-life-depends-water-climate-change-pollution-dams-threaten-iraqs-marsh-2021-10-14/.

NOTES TO THE AFTERWORD

1 "How Iraq's 'Deep State' Must be Challenged," Amwaj, accessed May 12, 2023, https://amwaj.media/article/how-iraq-s-deep-state-must-be-challenged.

2 "The Disappearance of an Iraq Environmental Activist Raises Concern for Human Rights Defenders," Iraqi Office for Human Rights, accessed February 9, 2023, https://iohriq.org/113-.html.

3 "Iraq Environmental Activist Jassim al-Asadi Freed Two Weeks After Kidnapping," The National News, accessed February 17, 2023, https://

www.thenationalnews.com/mena/2023/02/16/iraq-environmental-activ-ist-jassim-al-asadi-freed-two-weeks-after-kidnapping/.

4   "Special Representative of the Secretary-General for Iraq Ms. Jeanine Hennis-Plasschaert Iraq Climate Conference," UN Iraq, accessed March 18, 2023, https://iraq.un.org/en/222927-special-representative-secre-tary-general-iraq-ms-jeanine-hennis-plasschaert-iraq-climate.

SELECTED BIBLIOGRAPHY

Akanda, Ali, Sarah Freeman, and Maria Placht. "The Tigris-Euphrates River Basin: Mediating a Path Towards Regional Water Stability," *Al Nakhlah*. Spring (2007): 1–12. https://ciaotest.cc.columbia.edu/olj/aln/aln_spring07/aln_spring07g.pdf. Accessed November 10, 2020.

Altinbilek, Dogan. "Development and Management of the Euphrates–Tigris Basin," *Journal of Water Resources Development* 20, no. 1 (2004): 15–33.

Alwash, Suzanne. *Eden Again: Hope in the Marshes of Iraq*. Fullerton, California: Tablet House Publishing, 2013.

Al-Ansari, Nadhir, Ammar A. Ali, and Sven Knutsson. "Present Conditions and Future Challenges of Water Resource Problems in Iraq," *Journal of Water Resource and Protection* 6, no. 6 (2014): 1066–1098.

Broadbent, Geoffrey. "The Ecology of the *Mudhif*." *WIT Transactions on Ecology and the Environment* 113 (online, 2008): 15–26. https://www.witpress.com/elibrary/wit-transactions-on-ecology-and-the-environment/113/19193. Accessed October 5, 2020.

Canada/Iraq Marshlands Initiative, 2010. Atlas of the Iraqi Marshes. University of Victoria, Victoria, BC, Canada.

Canada/Iraq Marshlands Initiative, 2010. Managing for Change. University of Victoria, Victoria, BC, Canada.

Cockburn, Andrew, and Patrick Cockburn. *Out of the Ashes: The Resurrection of Saddam Hussein*. New York: Harper Collins, 1999.

Dellapenna, Joseph W., and Gupta Joyeeta, eds. *The Evolution of the Law and Politics of Water*. New York: Springer Publishing, 2008.

Haidar, Haidar. "Hymns of Death." in *A Banquet for Seaweed*, translated by Allen Hibbard and Osama Esber. *Tadween Publishing*, May

3, 2017, https://tadweenpublishing.com/blogs/news/haidar-haid-ars-hymns-of-death-translation-by-tadween-contributor-osama-esber. Accessed December 1, 2020.

Human Rights Watch. *The Iraqi Government Assault on the Marsh Arabs.* Human Rights Watch, 2003. https://www.hrw.org/legacy/backgrounder/mena/marsharabs1.htm.

Jacobsen, Thorkild. "The Battle between Marduk and Tiamat." *Journal of the American Oriental Society* 88, no. 1 (1968): 104–108.

Karsh, Efraim. *The Iran–Iraq War: 1980–1988.* Oxford: Osprey Publishing, 2014.

Kramer, Samuel Noah. *The Sumerians: Their History, Culture, and Character.* Chicago: University of Chicago Press, 1963.

Kubba, Sam, and Mudhafar Salim. "The Wetlands Wildlife and Ecosystem." In *The Iraqi Marshes and the Marsh Arabs: The Ma'dan, Their Culture and the Environment,* edited by Sam Kubba, 118–145. Ithaca Press, 2010.

Maitland, Alexander. *Wilfred Thesiger: The Life of the Great Explorer.* New York: The Overlook Press, 2011.

Nakash, Yitzhak. *The Shi'is of Iraq.* Princeton: Princeton University Press, 1994.

Nature Iraq. *Key Biodiversity Areas of Iraq: Priority Sites for Conservation and Protection.* Fullerton, California: Tablet House Publishing, 2017.

New Eden Master Plan, vol. 1, book 4, *The Marshlands.* Baghdad: Iraq Ministries of Environment, Water Resources, and Municipalities and Public Works, 2006.

Nicholson, Emma, and Peter Clark, eds. *The Iraqi Marshlands: A Human and Environmental Study.* London: Politico's Publishing, 2002.

Ochsenschlager, Edward L. *Iraq's Marsh Arabs in the Garden of Eden.* Philadelphia: University of Pennsylvania Museum of Archeology and Anthropology, 2004.

Polk, William, R. *Understanding Iraq: The Whole Sweep of Iraqi History, from Genghis Khan's Mongols to the Ottoman Turks to the British Mandate to the American Occupation.* New York: Harper Perennial, 2005.

Razoux, Pierre. *The Iran–Iraq War.* Translated by Nicholas Elliott. Cambridge, Massachusetts: The Belknap Press of Harvard University Press, 2015.

Saeed, Mahmoud. *Saddam City.* Translated by Ahmad Sadri. London: Saqi Press, 2004.

Thesiger, Wilfred. *The Marsh Arabs*. London: Longmans, Green, & Co., 1964.

Tripp, Charles. *A History of Iraq*. Cambridge: Cambridge University Press, 2007.

United Nations Environment Programme (UNEP). Partow, Hassan. *The Mesopotamian Marshlands: Demise of an Ecosystem*. Early Warning and Assessment Technical Report, UNEP/DEWA/TR.01-3 Rev. 1. Division of Early Warning Assessment. Nairobi, Kenya, 2001.

Watson, Julia. *Lo—Tek: Design by Radical Indigenism*. Cologne: Taschen, 2020.

Young, Gavin. *Return to the Marshes: Life with the Marsh Arabs of Iraq*. London: HarperCollins, 1977.

INDEX